DEPARTMENT OF THE ARMY FIELD MANUAL

155-MM ASSAULT GUN M53

AND 8-INCH HOWITZER M55,

SELF-PROPELLED

FIELD MANUAL

By **DEPARTMENT OF THE ARMY**
JULY 1957

©2013 Periscope Film LLC
All Rights Reserved
ISBN#978-1-937684-51-8
www.PeriscopeFilm.com

DISCLAIMER:

This document is a reproduction of a text first published by the Department of the Army, Washington DC. All source material contained herein has been approved for public release and unlimited distribution by an agency of the US Government. Any US Government markings in this reproduction that indicate limited distribution or classified material have been superseded by downgrading instructions promulgated by an agency of the US government after the original publication of the document No US government agency is associated with the publication of this reproduction. This manual is sold for historic research purposes only, as an entertainment. It contains obsolete information and is not intended to be used as part of an actual training program. No book can substitute for proper training by an authorized instructor.

©2013 Periscope Film LLC
All Rights Reserved
ISBN#978-1-937684-51-8
www.PeriscopeFilm.com

FM 6-93

Field Manual	HEADQUARTERS,
No. 6-93	DEPARTMENT OF THE ARMY
	WASHINGTON 25, D. C., *3 July 1957*

155-MM GUN M53, SELF-PROPELLED AND 8-INCH HOWITZER M55, SELF-PROPELLED

			Paragraphs	Page
CHAPTER	1.	GENERAL	1-4	4
	2.	ORGANIZATION	5, 6	9
	3.	SECTION DRILL		
Section	I.	General	7, 8	11
	II.	Preliminary commands and formations	9-14	12
CHAPTER	4.	PREPARING THE PIECE FOR FIRING AND TRAVELING		
Section	I.	Preparations for firing	15, 16	20
	II.	Preparations for traveling	17, 18	21
CHAPTER	5.	DUTIES IN FIRING, INDIRECT LAYING		
Section	I.	General	19, 20	22
	II.	Duties of chief of section	21-36	24
	III.	Duties of gunner	37-47	31
	IV.	Duties of cannoneers	48-95	37
CHAPTER	6.	FIRING BY DIRECT LAYING		
Section	I.	Technique of fire	96-102	55
	II.	Duties of chief of section	103-109	63
	III.	Duties of remainder of section	110-117	66

TAGO 7231C—July

		Paragraphs	Page
CHAPTER 7.	MISCELLANEOUS PROCEDURES AND TECHNIQUES	118–127	70
8.	BORE SIGHTING		
Section I.	General	128–132	81
II.	Testing target method	133–137	83
III.	Distant aiming point method	138, 139	87
IV.	Standard angle method	140–143	87
CHAPTER 9.	BASIC PERIODIC TESTS		
Section I.	General	144, 145	92
II.	Test of gunner's quadrant	146–150	94
III.	Test for telescope mount T179E2 and panoramic telescope T149E1	151–156	96
IV.	Fuze setters	157–159	99
CHAPTER 10.	MAINTENANCE AND INSPECTIONS	160–169	101
11.	DECONTAMINATION OF EQUIPMENT	170–173	106
12.	DESTRUCTION OF EQUIPMENT	174–177	109
13.	SAFETY PRECAUTIONS	178–182	111
14.	TRAINING	183–186	114
15.	TESTS FOR QUALIFICATION OF GUNNERS		
Section I.	General	187–194	117
II.	Test, direct laying, direct fire telescope	195–199	120
III.	Test, indirect laying, deflection only	200–204	122
IV.	Test, displacement correction	205–209	125
V.	Test, measuring deflection	210–214	127
VI.	Test, laying for elevation, elevation counter	215–219	128
VII.	Test, laying for elevation, gunner's quadrant	220–224	130

TAGO 7231C—July

		Paragraphs	Page
Section VIII.	Test, measuring elevation	225–229	131
IX.	Test, measuring angle of site to mask	230–234	132
X.	Test, sighting and fire control equipment	235–239	133
XI.	Test, materiel	240–244	136
APPENDIX	REFERENCES		139
INDEX			142

CHAPTER 1

GENERAL

1. Purpose and Scope

This manual is a guide to assist commanders in developing the sections of 155-mm gun M53 or 8-inch howitzer M55 firing batteries into efficient smooth-working teams that have a sense of discipline that will impel them to operate effectively under the stress of battle. This manual prescribes individual duties and section drills, tests and adjustments for sighting and fire control equipment, and instructions for the destruction and decontamination of equipment. The material contained herein is applicable, with modification, to both atomic and nonatomic warfare.

2. Definition of Terms

a. Section. Tables of organization and equipment (TOE's) prescribe the *personnel* and *equipment* comprising each section of a battery (figs. 1–3). In this manual the term *section* is often used to designate *only the personnel* required to serve the weapon and its equipment. TOE 6–437R and TOE 6–447R authorize the 5-ton truck as a section vehicle.

b. Front. The front of a section is the direction in which the muzzle of the gun (howitzer) points.

c. Right (Left). The direction right (left) is the right (left) of one facing to the front.

Figure 1. 155-mm gun M53, self-propelled, and section personnel.

d. In Battery. A gun (howitzer) is said to be in battery when the recoiling parts are in the normal firing position.

3. Description of Equipment

To insure proper use of the motor carriage and to avoid accidents caused by exceeding its capabilities

Figure 2. 8-inch howitzer M55, self-propelled, and section personnel.

Figure 3. A method of displaying section equipment.

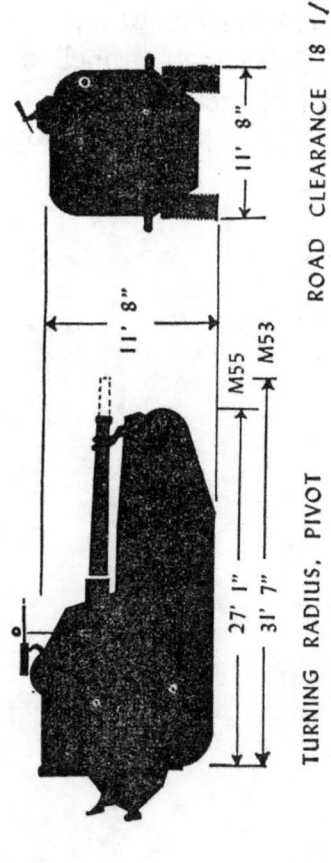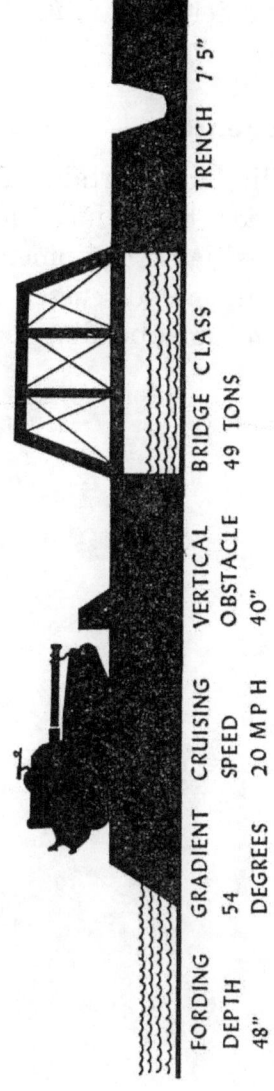

Figure 4. Performance characteristics of the motor carriage, 155-mm gun M53, self-propelled, or the 8-inch howitzer M55, self-propelled.

and limitations, all members of the section should be familiar with the performance characteristics shown in figure 4. For further details pertaining to full-track vehicle capabilities and combat driving, see TM 21–306.

4. References

Publications pertaining to the 155-mm gun M53, self-propelled; the 8-inch howitzer M55, self-propelled; and auxiliary equipment, covering related matters which are not discussed in detail in this manual, are listed in the appendix.

CHAPTER 2

ORGANIZATION

5. Composition of the Gun (Howitzer) Section

a. The gun (howitzer) section consists of section personnel; a 155-mm gun M53, self-propelled, or an 8-inch howitzer M55, self-propelled; a 5-ton truck; an ammunition trailer; and auxiliary equipment (figs. 1 and 2).

b. The personnel of the 155-mm gun section and the 8-inch howitzer section consists of—

 (1) A chief of section (CS).
 (2) A gunner (G).
 (3) Five cannoneers, numbered 1 through 5. No. 1 cannoneer is the assistant gunner.
 (4) An ammunition (ammo) specialist (AS).
 (5) Two ammunition handlers (AH).
 (6) Two drivers (D).

c. Section equipment is listed in TOE 6-437R for the 155-mm gun, TOE 6-447R for the 8-inch howitzer, and in appropriate standard nomenclature lists (SNL's) (see appendix).

6. General Duties of Personnel

a. Chief of Section. The chief of section is the noncommissioned officer in command of the section and, as such, is responsible to the battery executive for—

(1) Training and efficiency of personnel.

(2) Performance of duties listed under section drill, duties in firing, tests and adjustment of sighting and fire control equipment, and inspection and maintenance of all section equipment including the performance of scheduled preventive maintenance service on the motor carriage and section vehicle.

(3) Observance of safety precautions.

(4) Preparation of field fortifications for protection of equipment, ammunition, and personnel.

(5) Camouflage discipline; local security; and chemical, biological, and radiological warfare (CBR) security discipline.

(6) Maintenance of the weapon record book.

(7) Police of the section area.

b. Gunner. The gunner is the assistant to the chief of section in carrying out the duties specified in *a* above. The gunner's specific duties are prescribed in the appropriate chapters of this manual.

c. Cannoneers. Cannoneers perform duties as listed in this manual and any other duties that the chief of section prescribes.

d. Drivers. The drivers' primary duties are driving their respective motor carriage and performing preventive maintenance. They also perform such other duties as prescribed by this manual and by the training manuals pertaining to their motor carriages or as may be assigned by the chief of section. These duties can include substituting for any member of the section in firing.

CHAPTER 3

SECTION DRILL

SECTION I. GENERAL

7. Objective

The objective of section drill is the attainment of efficiency—maximum precision coupled with high speed.

8. Instructions

a. Adherence to drills prescribed in this manual is necessary to develop maximum efficiency and to prevent injury to personnel and damage to equipment. Section drill must be conducted in silence, except for commands and reports. The section must be drilled until reactions to commands are automatic, rapid, and efficient.

b. Errors are corrected immediately. Each member of the section must be impressed with the importance of reporting promptly to the chief of section any errors discovered before or after the command to fire has been given. The chief of section will report errors immediately to the executive.

c. Battery officers supervise the drill to insure that instructions are carried out and that maximum efficiency is obtained.

d. Duties should be rotated during training so that

each member of the gun (howitzer) section can perform all the duties within the section. In addition, battery overhead personnel not assigned specific duties during drill periods should be trained in the fundamentals of section drill in order that they will be capable of functioning efficiently with a gun (howitzer) section if required.

Section II. PRELIMINARY COMMANDS AND FORMATIONS

9. To Form the Section

a. To Fall In. The chief of section takes his post. On the command of execution the section forms in a double rank at close interval centered on and facing the chief of section at a distance of 3 paces (fig. 5). Higher numbered cannoneers, if present, form in order to the left of No. 5. The chief of section may indicate in his preparatory command the place and direction in which the section is to form. At the first formation for a drill or exercise, the caution, "As gun (howitzer) section(s)," precedes the command. The commands are FALL IN, or 1. IN FRONT (REAR) OF YOUR PIECE(S), 2. FALL IN, or 1. ON THE ROAD FACING THE PARK, 2. FALL IN. Execution is as follows: The section moves at double time and forms at close interval, at attention, guiding on the gunner. The driver of the 5-ton truck is to the left of the motor carriage driver and is the last in line. To execute 1. IN FRONT (REAR) OF YOUR PIECE, 2. FALL IN, the section falls in as shown in figure 5.

b. To Call Off. The section being in formation, the command is CALL OFF. At the command, all per-

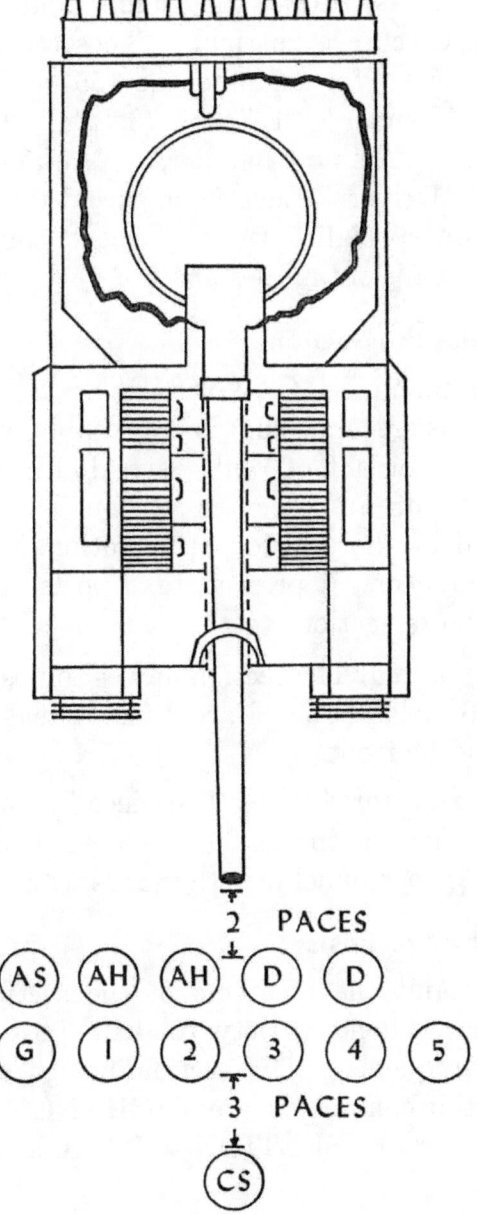

Figure 5. Formation of the section.

sonnel in ranks, except the gunner and ammunition specialist, execute eyes right. The section then calls off in sequence; for example, the 155-mm gun section calls off, "Gunner," "1," "2," "3," "4," "5," "Ammo specialist," "Ammo handler," "Ammo handler," "Driver," "Driver." Each man, except the gunner and ammunition specialist, turns his head smartly to the front as he calls out his designation.

10. To Post the Section

The command is 1. CANNONEERS, 2. POSTS. The command is general and is applicable whether the section is in or out of ranks, at a halt, or marching. All movements are executed at double time and are terminated at the position of attention. Higher numbered cannoneers, if present, take posts as prescribed by the chief of section.

a. Dismounted. The section moves to posts as shown in figure 6. All personnel are 2 feet outside the tracks and facing the front.

b. Prepared for Action. The piece having been prepared for action, the section is posted as shown in figure 7. All personnel face to the front.

11. To Change Posts

To acquaint the members of the section with all duties and to lend variety to drill, posts should be changed frequently. The section being *in formation* (fig. 5), the commands are 1. CHANGE POSTS, 2. MARCH, or 1. SECTION CHANGE POSTS, 2. MARCH.

a. At the command 1. CHANGE POSTS, 2. MARCH, all numbered cannoneers except No. 5 take

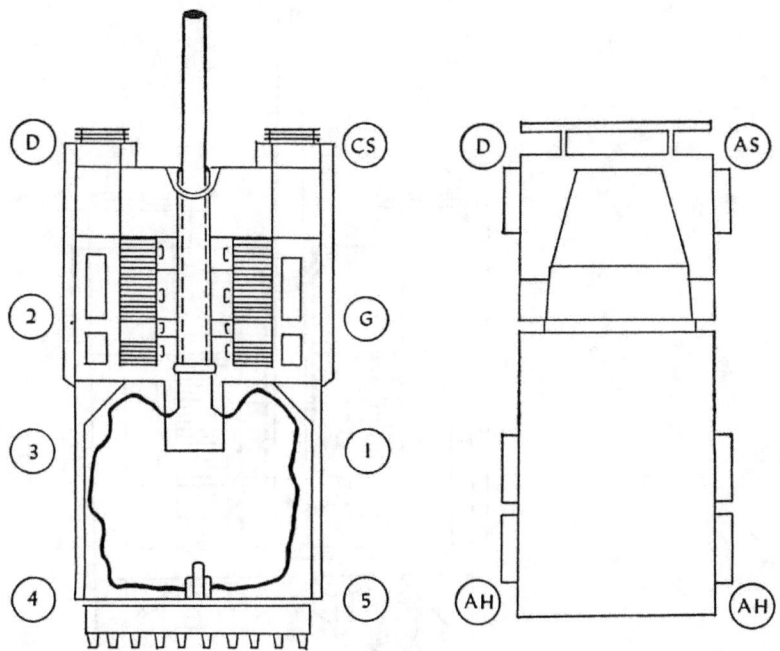

Figure 6. Posts of section dismounted.

two left steps, taking the position of the next higher numbered cannoneer. At the same time No. 5 moves at double time in rear of the front rank to the post of No. 1. All other members of the section stand fast.

b. At the command 1. SECTION CHANGE POSTS, 2. MARCH, all members of the section except No. 5 (or leftmost man in the front rank) and the driver of the 5-ton truck (or the leftmost man in the rear rank) take 2 left steps. No. 5 and the driver of the 5-ton truck move at double time in rear of the section and take the posts of the ammunition specialist and the gunner, respectively.

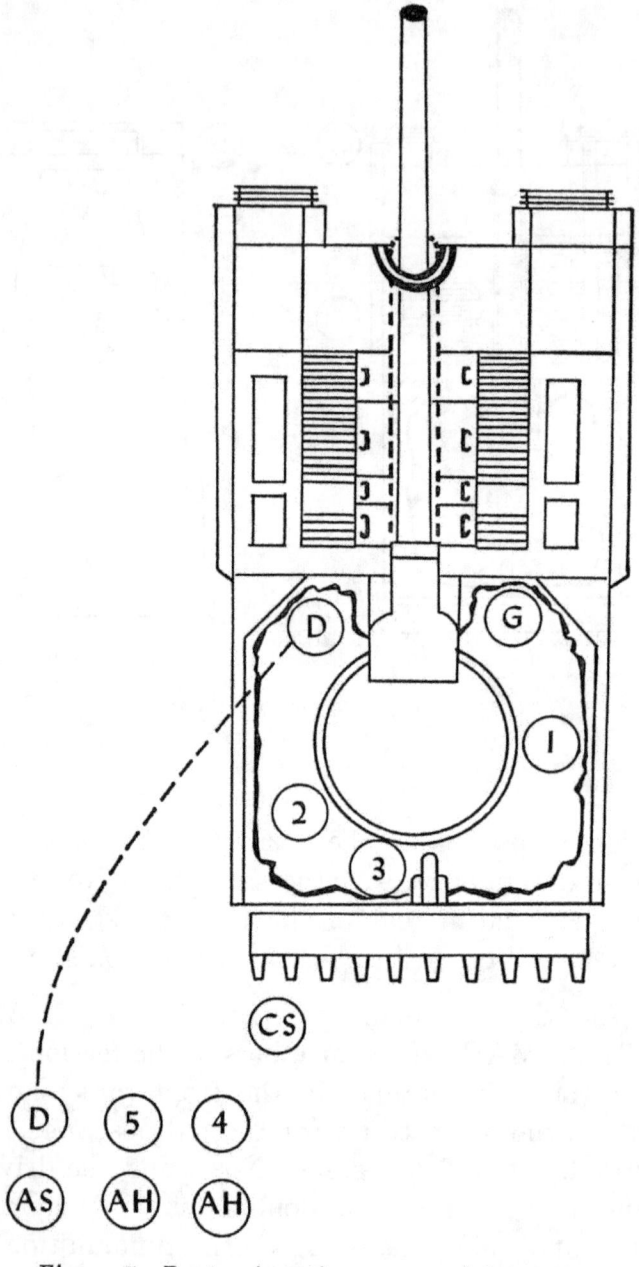

Figure 7. Posts of section prepared for action.

12. To Mount

The commands are 1. PREPARE TO MOUNT, 2. MOUNT, or MOUNT.

a. At the preparatory command, the section moves at double time to the positions shown in figure 6. At the command of execution, all personnel mount and take seats as indicated in figure 8. The chief of section, driver, gunner, and No. 1, 2, and 3 mount into the gun-howitzer motor carriage. No. 2, 3, and the driver, in that order, mount through the driver's door. No. 1, the chief of section, and the gunner, in that order, mount through the gunner's door. Likewise, at the command of execution, the ammunition specialist, driver, No. 4, 5, and the 2 ammunition handlers all mount at the same time into the section vehicle in the order shown in figure 6. If any members of the section are not to mount, their designation is announced with the caution, "Stand fast," given between the preparatory command and the command of execution. For example, 1. PREPARE TO MOUNT, DRIVERS STAND FAST, 2. MOUNT.

b. If the command is MOUNT, the section mounts in the manner and order prescribed for the command 1. PREPARE TO MOUNT, 2. MOUNT. Dismounted posts are not taken.

13. To Dismount

The commands are 1. PREPARE TO DISMOUNT, 2. DISMOUNT, or DISMOUNT.

a. At the preparatory command, the personnel mounted in the front seats of the weapon carriages unlatch and open the doors; members of the section

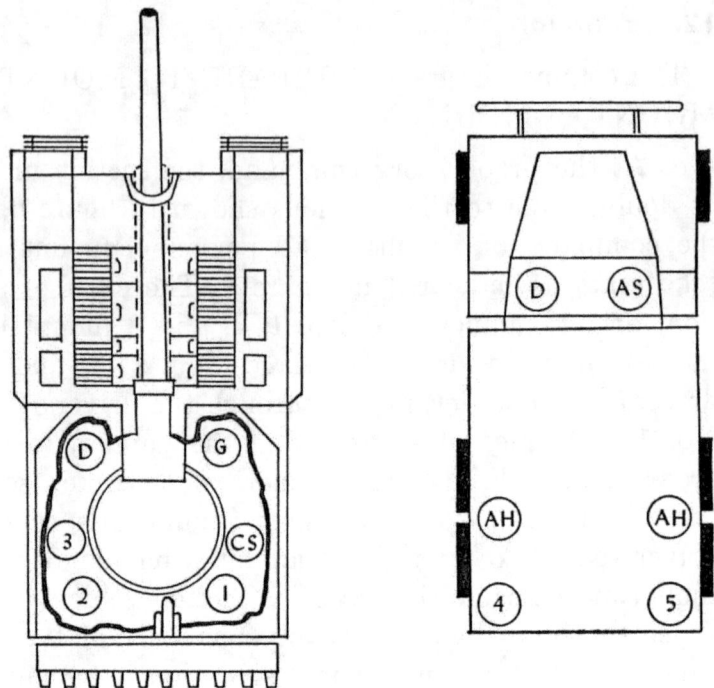

Figure 8. Section mounted.

assume positions from which they can dismount promptly. At the command of execution, they dismount and take their posts at double time as shown in figure 6.

b. If the command is simply DISMOUNT, the section executes all that is prescribed for the command 1. PREPARE TO DISMOUNT, 2. DISMOUNT.

14. To Fall Out

a. At Drill. When it is desired to give the personnel a rest from drill or to relieve them temporarily from *formation* or *post*, the command FALL OUT is given.

The command may be given at any time and means that the section is to remain in the drill area.

b. When Firing. When firing has been suspended temporarily, but it is desired to have the section remain in the vicinity of the motor carriage, the command FALL OUT is given. Men stand clear of the piece to insure that settings and laying remain undisturbed. During these periods, the chief of section may direct the men to improve the position, to replenish ammunition, or to do other necessary work.

CHAPTER 4

PREPARING THE PIECE FOR FIRING AND TRAVELING

Section I. PREPARATIONS FOR FIRING

15. General

The weapons of a battery will ordinarily be put into position individually under the direction of the executive and chiefs of section. A stake should be driven into the ground at a point where the center of each carriage is to be placed. Another stake should be placed in the direction of fire, 50 to 100 yards from the first stake, so that the driver of the motor carriage can point the tube at the far stake as he drives the vehicle into position over the first stake. Each vehicle is halted at its proper place by the chief of section. Hand signals for guiding the vehicle are found in FM 21-60, FM 25-10, and TM 21-306.

16. To Prepare for Action

a. The piece being in position or approaching it, the command is PREPARE FOR ACTION. Duties of individuals are given in table I. Each man takes his post (fig. 7) upon completion of his duties.

b. The piece normally will be partially prepared for action before reaching the firing position. The duties of the cannoneers in preparing for action are the same whether the piece is moving or halted, but only such operations as are practicable are carried out while moving. Immediately after the piece is established in position, preparation for action is completed without further command.

c. If PREPARE FOR ACTION has not been ordered before the piece is established in position, the command is habitually given by the chief of section as soon as the vehicle is halted in position. If preparation for action is not desired, the caution "Do not prepare for action" must be given.

Section II. PREPARATIONS FOR TRAVELING

17. March Order

To prepare to resume travel, the command is MARCH ORDER. Duties of individuals are given in table II. Each man takes his post (fig. 6) upon completion of his duties.

18. To Resume Firing in Another Position

a. If the piece is to be moved a short distance from where firing must be resumed promptly, the command MARCH ORDER is not given. When such a displacement is ordered, only those operations necessary for the movement of the motor carriage and the security of equipment are performed.

b. If the command MARCH ORDER is given while the weapon is prepared for travel as in *a* above, the operations pertaining to march order are completed.

CHAPTER 5

DUTIES IN FIRING, INDIRECT LAYING

Section I. GENERAL

19. Instructions

The general instructions in paragraphs 7 and 8 on the conduct of section drill apply equally to section drill in duties in firing by indirect laying. The sequence of duties performed in firing is shown in table III. For duties of the battery executive, see FM 6-40 and FM 6-140.

20. Duties of Individuals

In general, the duties of individuals in the section in indirect fire are given in *a* through *h* below. These duties apply to personnel of both the 155-mm gun and the 8-inch howitzer sections, unless specifically shown for one section or the other.

a. The chief of section supervises and commands his section and is responsible that all duties of the section are performed properly, all commands executed, and all safety precautions observed.

b. The gunner sets the announced deflection and elevation, lays for direction and elevation, refers the piece, sets the horizontal equilibrator, activates and deactivates the power traversing and elevating mechanism, and fires the piece electrically.

c. No. 1 opens and closes the breech, inspects chamber and bore, assists No. 2 in receiving projectile from No. 4 and 5 (155-mm gun), operates rammer controls, and activates safety fire switch.

d. No. 2 operates ammunition hoist (8-inch howitzer) or assists No. 1 in receiving projectile from No. 4 and 5 (155-mm gun), places projectile on loading trough (155-mm gun), inserts propellant into chamber, inserts primer, and removes spent primer.

e. No. 3 lowers and raises loading trough, swabs powder chamber after each round, and cleans obturator spindle vent, as required.

f. No. 4 fuzes projectiles, sets fuzes, and connects shot tongs to projectile (8-inch howitzer) or assists No. 5 in raising projectile to hand to No. 1 and 2 (155-mm gun).

g. No. 5 prepares propellant, cuts charges, and assists No. 4 in raising projectile to hand to No. 1 and 2 (155-mm gun).

h. The ammunition specialist is responsible for the care and stowage of ammunition. He also supervises handling and preparation of ammunition for firing.

i. The driver, motor carriage, shifts the carriage and assists in the preparation of ammunition as directed by the chief of section. He also starts and stops the auxiliary generator.

j. The driver, 5-ton truck, after his vehicle is unloaded, drives to the truck park, where he performs preventive maintenance operations unless directed otherwise by the ammunition specialist.

k. The ammunition handlers assist in the preparation of ammunition as directed by the ammunition specialist.

Section II. DUTIES OF CHIEF OF SECTION

21. List of Duties

(Detailed description of duties, par. 22–36.)

a. Indicates the aiming point to the gunner.

b. Measures the angle of site to the mask.

c. Follows fire commands.

d. Indicates when the piece is ready to be fired.

e. Gives the command to fire.

f. Reports errors and other unusual incidents of fire to the executive.

g. Records basic data.

h. Lays the piece for elevation, assisted by the gunner when the gunner's quadrant is used.

i. Measures the elevation.

j. Conducts prearranged fires.

k. Observes and checks frequently the functioning of the materiel.

l. Assigns duties when firing with reduced personnel.

m. Verifies the adjustment of the sighting and fire control equipment.

n. Controls the movement of the motor carriage.

o. Checks, before it is restowed for traveling, all ammunition not fired that has been prepared for firing.

22. Indicates the Aiming Point to the Gunner

When an aiming point has been designated by the executive (FM 6-140), the chief of section will make sure that he has properly identified the point designated. He will then indicate this point to the gunner. If there is any possibility of misunderstanding, the chief of section will turn the panoramic telescope until the horizontal and vertical cross hairs are on the aiming point designated.

23. Measures the Angle of Site to the Mask

a. The command is MEASURE THE ANGLE OF SITE TO THE MASK. The chief of section, sighting along the lowest element of the bore, directs the gunner to traverse and elevate the tube until the line of sight just clears the crest at its highest point in the probable field of fire. The gunner then turns the elevation counter knob in the appropriate direction until the three elevation vernier indexes are alined. The chief of section reads the elevation from the elevation counter and reports to the executive, "Sir, No. (so-and-so), angle of site (so much)."

b. When the executive announces the minimum elevation and charge or the minimum elevation for each charge, the chief of section records the data in a notebook and directs No. 1 to chalk the information on a convenient place on the hull or on the section data board (par. 127).

24. Follows Fire Commands

The chief of section will follow fire commands. He will repeat the commands as required.

25. Indicates When the Piece is Ready to be Fired

When the executive can see arm signals made by the chief of section, the chief of section will raise his right arm vertically as a signal that the piece is ready to be fired. He gives the signal as soon as the gunner calls "Ready." When arm signals cannot be seen, the chief of section reports orally to the executive, "Sir, No. (so-and-so) ready."

26. Gives the Command to Fire

When the gunner can see arm signals made by the chief of section, the chief of section will give the command to fire by dropping his right arm sharply to his side. When his arm signals cannot be seen, he commands orally NO. (SO-AND-SO), FIRE. The chief of section will not give the signal or command to fire until all cannoneers are in their proper places.

27. Reports Errors and Other Unusual Incidents of Fire to the Executive

If the piece cannot be fired, the chief of section will promptly report that fact to the executive and the reasons therefor; for example, "Sir, No. (so-and-so) out; misfire." When it is discovered that the piece has been fired with an error in laying, the chief of section will report that fact at once; for example, "Sir, No. (so-and-so) fired 40 mils right." When the gunner reports that the aiming posts are out of alinement, the chief of section will report that fact and, during the next lull in firing, ask permission to realine them. Likewise, the chief of section promptly reports other unusual incidents that affect the service of the piece.

28. Records Basic Data

The chief of section will record data of a semipermanent nature in a notebook. This includes such information as minimum elevations; aiming points used and their deflections; prearranged fires when section data sheets are not furnished; safety limits in elevation and deflection; number of rounds fired, with the date and hour; and calibration and special corrections when appropriate.

29. Lays the Piece for Elevation When Gunner's Quadrant is Used

a. Although the normal method of laying for elevation is by use of the elevation counter, the gunner's quadrant will be used to lay for elevation in the event of a malfunction of the elevation counter. The gunner's quadrant is also used to check the accuracy of the elevation counter. The command is QUADRANT (SO MUCH).

b. An elevation of quadrant 361.8, for example, is set on the gunner's quadrant as follows: the upper edge of the plunger plate is set opposite the 360 mark of the scale on the quadrant frame and the micrometer on the arm is turned to read 1.8. Care must be taken to use the same side of the quadrant in setting both the plunger plate and the micrometer.

c. The announced elevation having been set on the gunner's quadrant, the piece loaded, and the breechblock closed, the gunner's quadrant is set on the leveling plates of the breech ring. The words *line of fire* must be at the bottom of the quadrant with the arrow pointing toward the muzzle. The chief of section must be sure to use the arrow which appears on the same side of the quadrant as the scale which he is using. He stands squarely opposite the side of the quadrant and holds it firmly on the leveling plates parallel to the axis of the bore. *It is important that he take the same position and hold the quadrant in the same manner for each subsequent setting, so that in each case he will view the quadrant bubble from the same angle.*

d. The chief of section then directs the gunner to elevate or depress the tube until the bubble is centered, being careful that the last motion is in the direction in which it is more difficult to turn the handwheel. The chief of section cautions the gunner when the bubble is approaching the center, in order that the final centering may be performed accurately.

e. Normally, special and calibration corrections will be added algebraically at the battery fire direction center. The quadrant then would be announced as NO. (SO-AND-SO), QUADRANT (SO MUCH).

30. Measures the Elevation

At the command MEASURE THE ELEVATION, the piece having been laid, the chief of section directs the gunner to check the leveling of the telescope mount. The chief of section then sets the micrometer of the gunner's quadrant at zero and places it on the leveling plates of the breech ring. He then performs the following:

a. Moves the plunger arm of the gunner's quadrant until the bubble passes to the end of the vial away from plunger arm hinge.

b. Lowers the plunger arm slowly until the bubble just passes to the end of the vial toward the hinge.

c. Turns the micrometer until the bubble is accurately centered.

d. Removes the quadrant and reports the elevation thus set to the nearest 0.1 mil as "Sir, No. (so-and-so), elevation (so much)."

31. Conducts Prearranged Fires

When the execution of prearranged fires is ordered, the chief of section will conduct the fire of his section in conformity with the prescribed data.

32. Observes and Checks Functioning of the Materiel

The chief of section closely observes the functioning of all parts of the materiel during firing. Before the piece is fired, he makes sure that the recoil and counterrecoil systems contain the proper amount of oil; thereafter he carefully observes the functioning of these systems. He promptly reports to the executive any evidence of malfunctioning (TM 9–7212 and TM 9–7220).

33. Assigns Duties When Firing with Reduced Personnel

When the number of personnel serving the piece is temporarily reduced below that indicated in this manual, the chief of section will make such redistribution of duties as will best facilitate the service of the piece. Loss of cadremen, various details, and casualties will necessitate the section's operating with a reduced number of personnel to the extent that it is almost normal for section members to double up on duties. Around-the-clock firing will require the chief of section to divide the section into shifts to provide for relief.

34. Verifies the Adjustment of the Sighting and Fire Control Equipment

See TM 9–7212 (155-mm gun) or TM 9–7220 (8-inch howitzer) for detailed instructions on testing and adjusting sighting and fire control equipment.

35. Controls the Movement of the Motor Carriage

When it is necessary to move the motor carriage, the chief of section instructs the driver to start the engine. He then controls the displacement of the motor carriage by hand signals or by oral instructions. To shift the carriage when a new direction of fire is designated, the motor carriage should be moved so that when the tube is pointed in the new direction and the spade is seated the panoramic telescope will be over its original position and the aiming posts will still be in alinement.

36. Checks, Before it is Restored for Traveling, all Ammunition Not Fired That Has Been Prepared for Firing

The chief of section personally checks all ammunition not fired that has been prepared for firing (par. 77) before it is replaced in containers. He sees that powder increments prepared for firing are present in proper condition, are of the same lot number as the container, and are assembled in proper numerical order. He checks all time fuzes that have been set to see that they are reset to SAFE and that eyebolt lifting plugs are reinstalled. The chief of section also insures that grommets are replaced on the rotating bands of projectiles.

Section III. DUTIES OF GUNNER

37. List of Duties

(Detailed description of duties, par. 38–47.)

a. Centers the level bubbles on the panoramic telescope mount.

b. Lays the piece for direction.

c. Alines the aiming posts, assisted by No. 5.

d. Lays the piece for elevation.

e. Sets a common deflection to a common aiming point after the piece has been laid.

f. Sets or changes the deflection.

g. Signals and/or calls "Ready."

h. Refers the piece.

i. Makes corrections for aiming post displacement.

j. Fires the piece electrically.

38. Centers the Level Bubbles on the Panoramic Telescope Mount
(fig. 9)

The gunner centers the level bubbles on the telescope mount by operating the leveling knobs as part of preparing the sight for action. Once the bubbles are leveled, it should not be necessary to make further adjustments during firing. However, the level of the mount is verified before calling "Ready."

39. Lays the Piece for Direction
(fig. 10)

The piece being in position, tube in center of traverse, and not laid for direction, the executive com-

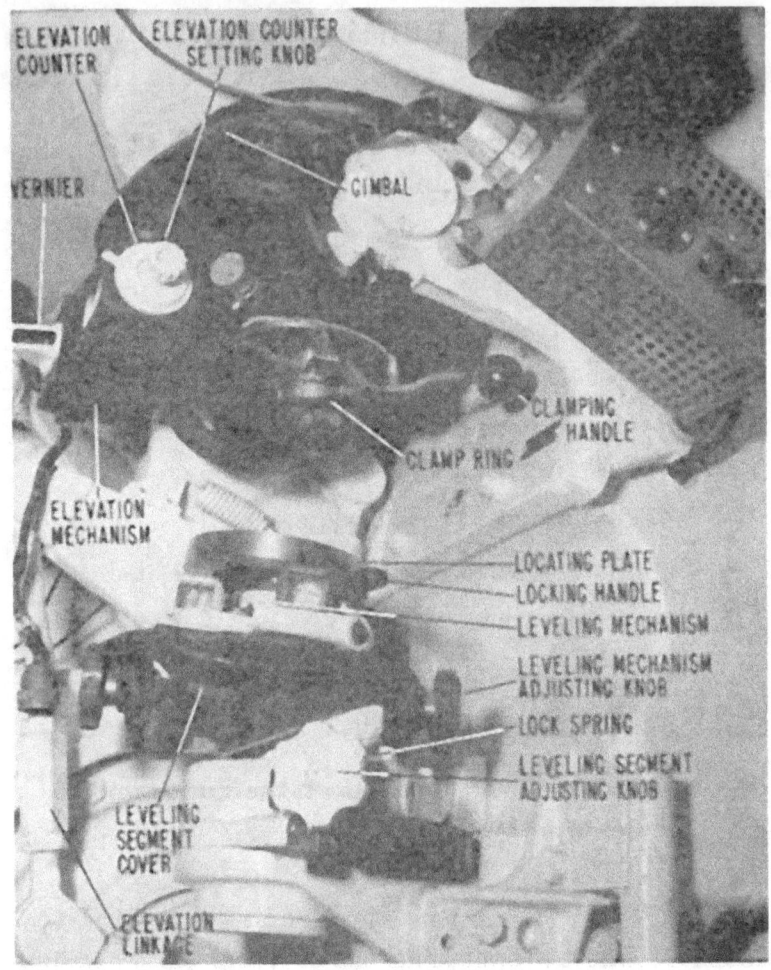

Figure 9. Telescope mount T179E2.

mands NO. (SO-AND-SO) ADJUST, AIMING POINT THIS INSTRUMENT. After the gunner has reported "Sir, No. (so-and-so) aiming point identified," the executive commands NO. (SO-AND-SO) DEFLECTION (SO MUCH). The gunner sets the commanded deflection on the coarse azimuth scale, as viewed through the window, and the micrometer scale. He then directs the driver to move the motor carriage until the vertical cross hair of the sight reticle is approximately on the executive's aiming circle. The gunner then traverses the tube until the vertical cross hair of the sight is exactly centered on the executive's aiming circle. He checks to insure that the bubbles are level and announces "Sir, No. (so-and-so) ready for recheck." As additional deflections are announced by the executive, he sets them on the coarse azimuth and the micrometer scales and traverses the tube so that the vertical line of the sight is centered on the aiming circle. When the executive announces "No. (so-and-so) is laid," the tube is oriented and should not be traversed except on order of the executive.

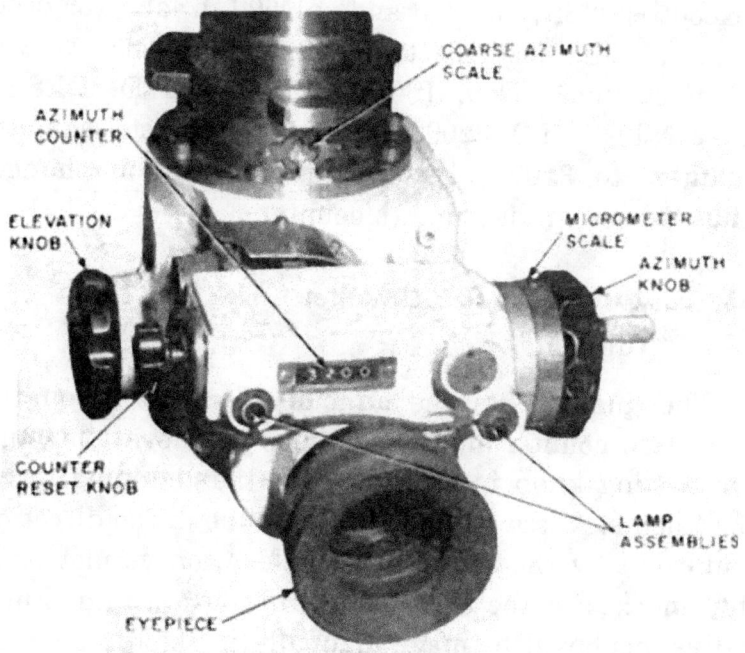

Figure 10. Panoramic telescope T149E1.

40. Alines the Aiming Posts, Assisted by No. 5

The piece having been laid as in paragraph 39, the executive may command AIMING POINT, AIMING POSTS, DEFLECTION 600, REFER. At the command the gunner sets 6 on the coarse azimuth scale and 0 on the micrometer scale and, using hand signals, directs No. 5 in the alinement of the aiming posts with the vertical line of the sight reticle. If, because of the nature of the terrain, the aiming posts cannot be set at deflection 600, the gunner turns the micrometer knob until the coarse azimuth and micrometer scales are on another even-hundred-mill graduation. He alines the aiming posts at this new deflection. The chief of section reports the altered deflection to the executive, "Sir, No. (so-and-so) 600 in lake (or other reason), laid on deflection 800." The executive will then command NO. (SO-AND-SO) LAID, RESET COUNTER TO 3200. The gunner resets azimuth counter to 3200. All subsequent deflection changes must be set on the azimuth counter.

41. Lays the Piece for Elevation
(fig. 11)

The gunner sets the announced elevation on the elevation counter and then clamps the elevation counter setting knob in position with the clamping knob. He then elevates or depresses the tube in the direction indicated by the arrows in the elevation vernier until the indexes in the elevation vernier are alined. Final adjustment will be made manually in the direction of increasing resistance.

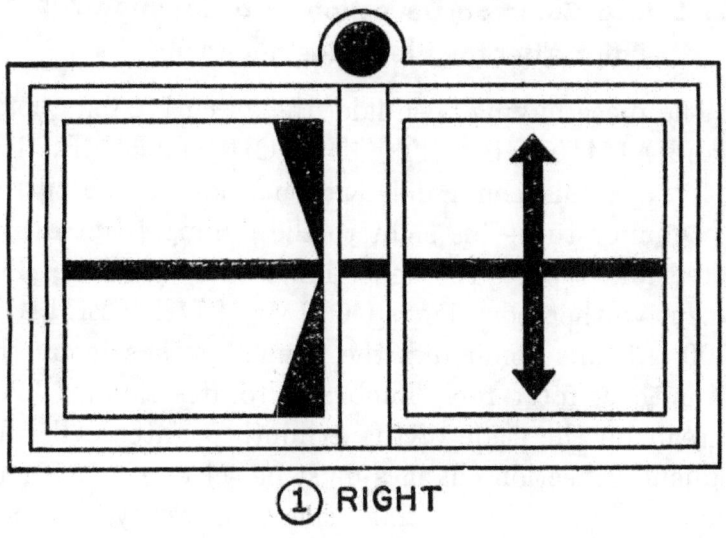

① RIGHT

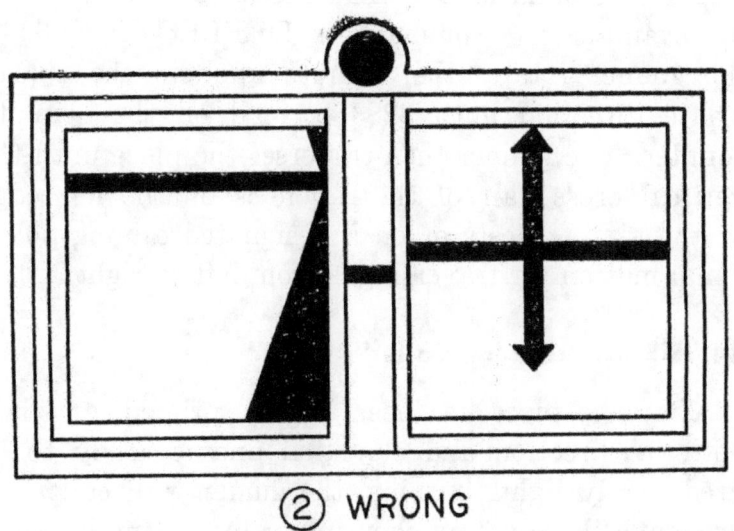

② WRONG

Figure 11. Alining elevation vernier indexes.

42. Sets a Common Deflection to a Common Aiming Point After the Piece Has Been Laid

The piece having been laid, the executive may command AIMING POINT, CHURCH STEEPLE, REFER. At this command, without moving the turret, the gunner turns his sight to the aiming point designated and reports the deflection to the executive. The executive then commands COMMON DEFLECTION 3200. At this command, the gunner pushes in on the azimuth counter reset knob and rotates it until 3200 is read on the scale of the azimuth counter. All subsequent deflection changes must be set on the azimuth counter. He makes a final check to verify that the line of sight is still on the aiming point.

43. Sets or Changes the Deflection

The command is DEFLECTION (SO MUCH). If, for example, the command is DEFLECTION 3283, the gunner rotates the azimuth knob in the appropriate direction until 3283 is read on the azimuth counter. The gunner then traverses the piece until the vertical cross hair of the reticle is on the left edge of the aiming posts or on a designated aiming point. Final motion for traversing is from left to right.

44. Signals and/or Calls "Ready"

After the piece has been loaded, primed, and laid both in direction and elevation and is ready to be fired, ready lights burning, the gunner will call and/or signal "Ready," by shouting or by raising his right arm to signify that the piece is ready to be fired.

45. Refers the Piece

The command from the executive is AIMING POINT THIS INSTRUMENT (OR OTHER POINT), REFER. Without disturbing the laying of the piece, the gunner turns only the sight until, with bubbles level, the vertical cross hair of the reticle is on the point designated. He then reports the deflection as read on the coarse azimuth and the micrometer scales to the executive, "Sir, No. (so-and-so) deflection (so much)."

46. Makes Corrections for Aiming Post Displacement

For details of correcting for aiming post displacement, see paragraph 121.

47. Fires the Piece Electrically

At the chief of section's signal or command, the gunner steps on the foot firing switch to fire the piece electrically. For information on firing the piece manually, see paragraph 61.

Section IV. DUTIES OF CANNONEERS

48. No. 1, List of Duties

(Detailed description of duties, par. 49–52.)
a. Opens and closes the breech.
b. Receives projectiles.
c. Rams projectiles.
d. Operates safety fire switch.

49. Opens and Closes the Breech

After the piece has been fired (after lanyard is removed in manual firing), No. 1 disengages the breechblock operating lever catch. He pulls the breechblock operating lever downward to the horizontal and turns the operating lever in an arc to the right until the breech is fully open. He then inspects the chamber and bore; if clear, he announces "Bore clear." To close the breech, the above process is reversed.

50. Receives Projectiles (155-MM Gun Only)

Assisted by No. 2, No. 1, standing on the lower rear turret door, bends down and grasps the right carrying handles of the loading tray as it is lifted by No. 4 and 5. No. 1 and 2 set the loading tray on the rear lower turret door to the left of the rammer-spade hoist motor housing.

51. Rams Projectiles

After a projectile is placed on the lower loading trough against the rammer chain head, the chief of section will signal No. 1 to ram the projectile by making a horizontal sweeping motion with his left arm in the direction of the breech. On receipt of this signal, No. 1 will rapidly move the rammer power selector lever into the RAM position. As soon as the projectile is firmly seated, No. 1 will rapidly move the selector lever to the RETURN position to retract the rammer chain.

52. Operates Safety Fire Switch

After No. 2 inserts the primer in the firing lock, No. 1 pushes the red safety fire switch, located on the right trunnion, to the ON position. He makes a check to see that the light, located on the auxiliary panel, is burning.

Caution: The piece is now ready to be fired electrically.

53. No. 2, List of Duties

(Detailed description of duties, par. 54–61.)

a. Operates the ammunition hoist (8-inch howitzer).

b. Assists No. 1 in receiving the projectile (155-mm gun).

c. Places the projectile on the loading trough (155-mm gun).

d. Inserts the propellant into the chamber.

e. Inserts the primer into the firing lock.

f. Removes spent primer.

g. Prepares the firing lock for manual firing.

h. Fires the piece manually.

54. Operates the Ammunition Hoist (8-Inch Howitzer)

After insuring that the front track and the rear tract are locked in alinement, No. 2 operates the ammunition hoist controls to move the hoist down to a level where No. 4 can easily reach them. No. 4 relieves No. 2 in the operation of the hoist and connects the shot tongs of the hoist to the fuzed projectile. After the projectile is connected to the shot tongs, No. 4 raises the ammunition hoist to a level where No. 2 can relieve him. After No. 2 relieves No. 4 of the controls, he moves the hoist so that the projectile is positioned on the loading trough with its base against the rammer chain head. No. 2 disengages the shot tongs and raises the hoist vertically to move it out of the way. The ammunition hoist must not be on the rear track when the piece is fired.

55. Assists No. 1 in Receiving the Projectile (155-MM Gun)

No. 2, standing on the lower rear turret door, grasps the left carrying handles of the loading tray as it is raised by No. 4 and 5. No. 2 and No. 1 then place the loading tray, with projectile, on the lower rear turret door to the left of the rammer-spade hoist motor housing.

56. Places the Projectile on the Loading Trough (155-MM Gun)

No. 2 picks up the projectile from the loading tray after the loading tray and projectile are placed on the lower rear turret door and places it in the loading trough with the base of the projectile positioned against the rammer chain head.

57. Inserts the Propellant into the Chamber

As the projectile is rammed by No. 1, No. 2 receives the prepared powder charge from No. 3. No. 2 then inserts the powder charge into the breech from the left so that the igniter pad is 3 inches within the chamber.

58. Inserts the Primer into the Firing Lock

After the breech is closed, No. 2 grasps the firing lock hammer with his right hand and pulls it out approximately ¼ inch. He rotates it counterclockwise until the wedge uncovers the primer chamber. He inserts a primer and rotates the hammer in a clockwise direction until it stops in a vertical position with the wedge fully raised.

59. Removes Spent Primer

After the piece has been fired, No. 2 removes the spent primer from the firing lock by turning the operating handle in a counterclockwise direction.

60. Prepares the Firing Lock for Manual Firing

The primer being inserted in the firing lock as described in paragraph 58, No. 2 rotates the hammer in a clockwise direction while exerting a rearward pressure. He continues to turn the hammer until the hammer is retained in the cocked position. No. 1 does not place the safety switch in the FIRE position until directed to do so by the chief of section.

61. Fires the Piece Manually

When directed by the chief of section, No. 2 prepares the firing lock for mechanical operation by cocking the firing lock, attaching a lanyard to the trigger, placing the safety lever in the fire position, and indicating to the chief of section that the piece is ready to be fired. At the chief of section's signal or command, No. 2 grasps the handle of the lanyard with his right hand and pulls strongly with a quick steady movement to the left rear. Immediately after firing, No. 2 detaches the lanyard. In case of a misfire, the instructions contained in paragraph 180 will be followed.

62. No. 3, List of Duties

(Detailed description of duties, par. 63–66.)
   a. Lowers the loading trough.

b. Raises the loading trough.

c. Cleans the obturator spindle vent.

d. Swabs the powder chamber and inspects the bore after each round.

63. Lowers the Loading Trough

After the breech has been opened at loading elevation, No. 3 grasps the loading trough handle and pulls it out to free the detent from the hole in the loading trough. He then pushes the handle forward and allows the loading trough to lower into its extended position.

64. Raises the Loading Trough

As soon as No. 2 has inserted the propellant into the chamber, No. 3 pulls out on the loading trough handle and lifts the loading trough into its stowed position, making certain that the detent in the handle seats itself in the hole in the trough.

65. Cleans the Obturator Spindle Vent

No. 3 cleans the primer vent when possible and during lulls in firing with the vent cleaning tools.

66. Swabs the Powder Chamber and Inspects the Bore After Each Round

After each round, No. 3 swabs out the powder chamber immediately after No. 1 opens the breech. The rear of the bore, including the forcing cone, is swabbed with a water-soaked swab which may be improvised by wrapping burlap around one end of a rammer staff section. After swabbing between rounds and before reloading the piece, No. 3 inspects the bore for damage to the tube, burning fragments of powder bags, or other objects. Any burning fragment or other objects in the bore must be removed before reloading. If bore is clear, No. 3 calls out "Bore clear." Any damage to the bore must be reported to the chief of section immediately.

67. No. 4, List of Duties

(Detailed description of duties, par. 68–74.)

a. Fuzes or changes fuzes of projectiles.

b. Sets the fuze setter M26 or M28.

c. Sets fuzes.

d. Removes fuzes from projectiles.

e. Assists No. 2 in operating the ammunition hoist (8-inch howitzer).

f. Raises the projectile on the loading tray (155-mm gun).

g. Assists in carrying the projectile (8-inch howitzer).

68. Fuzes or Changes Fuzes of Projectiles

No. 4 unscrews the eyebolt lifting plug from the fuze socket of the projectile; inspects the socket for rust and dirt; removes (or replaces) the supplementary charge, if necessary; and screws in the designated fuze. In tightening or loosening the fuze of the projectile, only the authorized fuze wrench should be used. Variable time (VT) fuzes should be screwed in by hand and tightened with fuze wrench M18 by using manual force only. *Do not hammer on the wrench or use an extension handle.* If a time fuze is used, No. 4 removes the safety pull wire from the fuze and, if a booster is present, the safety pin from the booster. Boosters without safety pins must not be used.

69. Sets the Fuze Setter M26 or M28

No. 4 releases the time scale clamping screw marked "T" and grasping the handle, turns the body until the index on the body is opposite the announced time on the time scale. He then locks the time scale clamping screw, being careful not to disturb the setting. For accuracy, he looks squarely at the scales and indexes in the same manner each time.

70. Sets Fuzes

a. Selective Superquick and Delay Fuze. When FUZE QUICK is announced, No. 4 will verify the superquick setting. (The slot on the setting sleeve should be alined with the letters SQ.) When FUZE DELAY is announced, No. 4 will turn the setting sleeve with a fuze wrench, screwdriver, or similar tool until the slot is alined with the word DELAY.

b. Combination Time and Superquick Fuze. The combination time and superquick fuze may be set for time action. However, the percussion element will detonate the round upon impact if the time element fails. After fuzing the projectile, No. 4 removes the safety pull wire from the fuze. For percussion action, the command is FUZE M55 QUICK. No. 4 then verifies that the S on the setting ring is alined with the index on the fixed ring; if not, he sets it at S.

c. Time Fuzes.

    (1) *Using fuze setter M26 or M28.* After making the announced settings on the fuze setter, No. 4 removes the safety pull wire from the

fuze, carefully places the fuze setter over the fuze, and turns the setter in the direction of increasing readings until the notch on the time ring of the fuze engages the stop on the setting ring of the fuze setter. He places the handle in the most convenient position, pushes down the fuze setter until the notch fully engages the stop, and continues to turn it in the direction of increasing readings until the pawl in the adjusting ring assembly drops into the notch of the fixed fuze ring. This action prevents further turning and indicates that the fuze is set. He then lifts the fuze setter from the fuze without rotating it and makes a visual check of the fuze setting to insure that the fuze ring notch was actually engaged and that the fuze is properly set. Once set, if the time setting on the fuze is to be changed, the fuze setter is reset to the desired time setting, and the fuze is set again as described above.

(2) *Using fuze setter M14 or M27.* Fuze setter M14 or M27 is a wrench-type fuze setter in which the fuze time scale is used in setting the fuze. After the safety pull wire has been removed, No. 4 places the fuze setter on the fuze with the taper contour of the hole fitting the fuze. He engages the key in the wrench with the slot on the fuze and turns the wrench in the direction of increasing readings until the index mark on the fuze alines with the required time setting on the fuze scale. No. 4 then removes the wrench,

being careful to avoid changing the setting, and makes visual check of the fuze to insure that the fuze is properly set.

d. VT Fuzes. The older type VT fuzes (M97) operate and function in such a manner as to require no setting by personnel. The new type VT fuzes (M514-series) have a time setting ring and are set by using the fuze setter M28 in the same manner as the M54-series time fuzes are set by using the fuze setter M26. However, if the fuze setter M28 is not available, the wrench-type fuze setter is used to set the new type VT fuzes. VT fuzes of certain lots are issued with a wax coating on the plastic ogive. This wax coating is necessary for the proper functioning of the fuze and should not be removed. VT fuzes should be used as issued; that is, with the wax coating on the ogive, if so issued, or without a wax coating, if so issued.

71. Removes Fuzes from Projectiles

If a projectile that has been fuzed is not to be fired, the fuze is removed. The operation of inserting a fuze is reversed. Supplementary charges will be replaced, provided the projectile was issued with the charge. The booster cotter pin of the fuze is replaced if a booster is used. Combination superquick and delay fuzes are reset to SQ (superquick). Time fuzes are reset to S (safe) and the safety pull wire replaced. M514-series VT fuzes are reset to initial setting as shipped; i.e., to S or O (variation depending on model number). All fuzes are returned to their containers. The eyebolt lifting plugs are replaced in the fuze sockets of the projectiles.

72. Assists No. 2 in Operating the Ammunition Hoist (8-Inch Howitzer)

When No. 2 lowers the ammunition hoist to a level where No. 4 can grasp the controls, No. 4 takes control of the hoist and secures the shot tongs to the projectile. He then raises the hoist with the projectile to a level where No. 2 can relieve him of the hoist controls.

73. Raises the Projectile on the Loading Tray (155-MM Gun)

After the projectile has been prepared for firing, No. 4 places it on the loading tray. He grasps the handles on the left side of the tray. No. 5 grasps the handles on the right, and together they carry the tray to the piece and hand it up to No. 1 and 2 (155-mm gun).

74. Assists in Carrying the Projectile (8-Inch Howitzer)

No. 4, assisted by No. 5, places a fuzed projectile on the loading tray. No. 4 takes the front left handle. No. 5 takes the front right handle, and the 2 ammunition handlers each take 1 of the rear handles. At the command LIFT by the ammunition specialist, all four men lift the loading tray. They carry the tray to the rear of the motor carriage and set it down under the ammunition hoist at the command LOWER by the ammunition specialist.

75. No. 5, List of Duties

(Detailed description of duties, par. 76–80.)

a. Sets out aiming posts.

b. Prepares powder charges.

c. Passes powder charge to No. 2.

d. Calls out the number of the charge.

e. Assists in carrying the projectile.

76. Sets Out Aiming Posts

No. 5 sets out aiming posts as described in paragraph 120.

77. Prepares Powder Charges

a. The propelling charge for the 155-mm gun is composed of a base charge and one increment. The base section only (increment section removed) is known as the *normal* charge and is used for all obtainable ranges. The full charge (base and increment) is known as the *supercharge* and is used for extreme ranges. (It may also be used in direct laying.) The rear end of the base section contains an igniter pad which is usually dyed red. A cup-shaped, felt-based cloth cover fastened to the charge by a drawstring protects the igniter pad and is removed just before loading. Four tying straps are sewed to the front end of the base section and serve to assemble base and increment sections into one propelling charge. When the command designating the charge is given, for example, NORMAL CHARGE, No. 5, assisted by the driver, when available, takes a complete charge from a container, places the complete charge in front of him, base charge on the bottom, and unties the straps which hold the bags together. He removes the increment charge from the top and then loosely and uniformly reties the straps. No. 5 then removes the igniter protector cap and the ammunition data tags. Unused increments are disposed of as directed by the battery executive. In night firing when the flash reducer M1 is to be used, No. 5 will tie the flash reducer around the propelling charge as described in TM 9–1901.

Caution: Containers should not be opened until just before the charges are to be used.

b. Two propelling charges are available for the 8-inch howitzer. These are charge, propelling, M1 (green bag) and charge, propelling, M2 (white bag). *Under no circumstances will sections of the green bag charge be mixed with sections of the white bag charge (TM 9–1901), and powder containers should not be opened until just before the charges are to be used.* The propelling charge M1 is composed of a base charge and 4 unequal increments corresponding to the first 5 zones of fire. As in *a* above, this charge contains an igniter pad, igniter pad protector, and four typing straps which must be removed and retied when decreased charges are used in firing. The base charge is charge 1 and the other charges are numbered from 2 to 5, inclusive. When the command designating the charge is given, for example, CHARGE 4, No. 5, assisted by the driver, when available, takes a complete charge from a container, places it in front of him, base charge on the bottom, and unties the straps which hold the bags together. He removes the bag marked "5" from the top, leaving the bag marked "4" at the top of the pile, and checks all charges to ascertain their presence in the proper order. He then loosely and uniformly reties the straps and then removes the cloth igniter pad protective cover. As this propelling charge is considered flashless under all conditions, no flash reducer is provided. Propelling charge M2 (white bag) is composed of a base charge and two increments. The base charge is the equivalent of charge 5. The increments provide charges 6 and 7. The preparation of a charge of this type is similar to that described, except that no charge lower than charge 5 can be prepared. In night firing when the flash reducer M3 (T3) is to be used, No. 5 will assemble the propelling charge and insert the flash reducer at the forward end under the tie straps.

78. Passes Powder Charge to No. 2

As the projectile is being seated in the breech and after the powder charge has been prepared as in paragraph 77, No. 5 passes the prepared charge to No. 3. He hands it in such a way that No. 2 is able to grasp the base with his right hand.

79. Calls Out the Number of the Charge

After passing the powder charge to No. 3, No. 5 calls out the number of the charge he has prepared; for example, "Normal charge" (155-mm gun) or "Charge 3" (8-inch howitzer). This informs the chief of section and the ammunition specialist that the proper charge has been prepared.

80. Assists in Carrying the Projectile

No. 5 will assist in carrying the projectile as described in paragraph 73 or 74, as appropriate.

81. Ammunition Specialist, List of Duties

(Detailed description of duties, par. 82–87.)

a. Receives and accounts for ammunition for the section.

b. Supervises ammunition handling.

c. Supervises the storage of ammunition.

d. Supervises the preparation of ammunition for firing.

e. Follows fire commands and insures that the designated powder charge, projectile, and fuze are used.

f. Determines powder temperature and announces it when so directed.

82. Receives and Accounts for Ammunition for the Section

The ammunition specialist receives and accounts for the ammunition as may be provided the section. He verifies the amount received and receipts for it, maintains a daily record of all ammunition received and fired, and keeps the chief of section informed as to the status of the ammunition supply within the section.

83. Supervises Ammunition Handling

The ammunition specialist requires the cannoneers to handle ammunition properly. He prevents any of the following:

a. Smoking in the vicinity of ammunition or the use of any lights, other than battery or generator powered lights, in the vicinity of powder charges.

b. Rough handling of ammunition, including dropping projectiles, powder containers, fuzes, and primers or allowing projectiles to strike together and ammunition to become dirty, wet, or overheated.

c. The arming delay time for VT fuzes from being set more than twice or being backed-up.

84. Supervises the Storage of Ammunition

For details regarding storage of ammunition, see paragraph 126, this manual, FM 6-140, FM 9-6, and TM 9-1900.

85. Supervises the Preparation of Ammunition for Firing

The ammunition specialist carefully supervises the work of the cannoneers in preparing rounds for firing. He sees that the projectiles are cleaned thoroughly and that all burrs on the rotating bands are removed by filing. He requires that all powder charges be kept in their closed containers until just before loading, and that fuzes be kept in their boxes until just before they are needed. He will insure that powder charges are properly segregated by lot number and that base charges and increment sections of different lot numbers do not become mixed.

86. Follows Fire Commands and Insures That the Designated Powder Charge, Projectile, and Fuze Are Used

The ammunition specialist follows the fire commands and indicates to the cannoneers concerning the projectile, powder charge, and fuze to be used. For any single firing mission, he sees that the projectiles are all of the same weight and that the powder charges and time fuzes are each of one lot number.

87. Determines Powder Temperature and Announces It When so Directed

Propellants must be protected from excessive and rapid changes in temperature. High temperatures greatly accelerate the normal rate of deterioration and cause excessive and irregular chamber pressures in firing, resulting in erratic ranges. Sudden changes in temperature may also cause moisture to condense upon the charges. The ammunition specialist determines powder temperature and announces it when so directed by the chief of section.

88. Ammunition Handlers (Two), List of Duties

(Detailed description of duties, par. 89, 90.)

a. Assist in carrying projectiles (8-inch howitzer).

b. Assist in preparation of ammunition.

89. Assist in Carrying Projectiles (8-Inch Howitzer)

See paragraph 74 for information on carrying ammunition.

90. Assist in Preparation of Ammunition

The ammunition handlers assist No. 4 and 5 in preparing ammunition as directed by the ammunition specialist.

91. Driver, Motor Carriage, List of Duties

(Detailed description of duties, par. 92-94.)

a. Shifts the carriage.

b. Performs maintenance in stabilized situations.

c. Assists No. 5 in preparing powder charges.

92. Shifts the Carriage

When a new section of fire is designated, the driver shifts the motor carriage as directed by the chief of section (par. 35).

93. Performs Maintenance in Stabilized Situations

When the situation is stabilized, the driver performs such preventive maintenance as may be accomplished without interfering with the firing of the piece. Any disassemblies or maintenance operations that will render the vehicle immobile for any period of time must be ordered by the chief of section.

94. Assists No. 5 in Preparing Powder Charges

As soon as his other duties permit or when directed by the chief of section, the motor carriage driver assists No. 5 in preparing powder charges (par. 77). He holds the complete charge while No. 5 prepares the charge. The driver then places any discarded powder increments in the containers provided for that purpose.

95. Driver, 5-Ton Truck

After the cargo vehicle has been unloaded, the driver moves it to the truck park or performs other duties as may be prescribed by the ammunition specialist. At the truck park, unless otherwise directed, the driver performs such preventive maintenance service as may be accomplished without making his vehicle immobile.

CHAPTER 6

FIRING BY DIRECT LAYING

Section I. TECHNIQUE OF FIRE

96. General

a. Firing by direct laying is a technique that demands special training. The section must operate as an independent unit. Training in direct laying is based on the technique employed in indirect laying. Targets taken under fire by the section in direct laying are usually those capable of returning direct fire on the weapon section; therefore, the speed and accuracy required in indirect laying becomes even more important for direct laying missions.

b. For additional information on firing by direct laying, see FM 6-40 and FM 6-140.

97. Preparation of a Range Card

a. The chief of section is responsible for defense in his assigned sector, but he should be prepared to fire on targets in other sectors.

b. As soon as possible after occupation of position, the chief of section measures or estimates the ranges to critical points in likely avenues of approach for enemy tanks and vehicles. For quick references, he prepares a range card (fig. 12) upon which he notes these ranges.

c. If there are no prominent terrain features, stakes may be driven into the ground at critical points for reference. As time permits, the range card is improved by replacing estimated ranges with more accurate ranges obtained by firing, pacing, taping, vehicle speedometer reading, map measurements, or survey.

98. Field of Fire

The field of fire for the piece should be cleared of all obstructions that might endanger battery personnel when the piece is fired or that might hinder observation. Care should be taken not to give away the location of the position.

99. Conduct of Fire

Targets for direct laying usually consist of vehicles, tanks, and personnel threatening the battery. Enemy

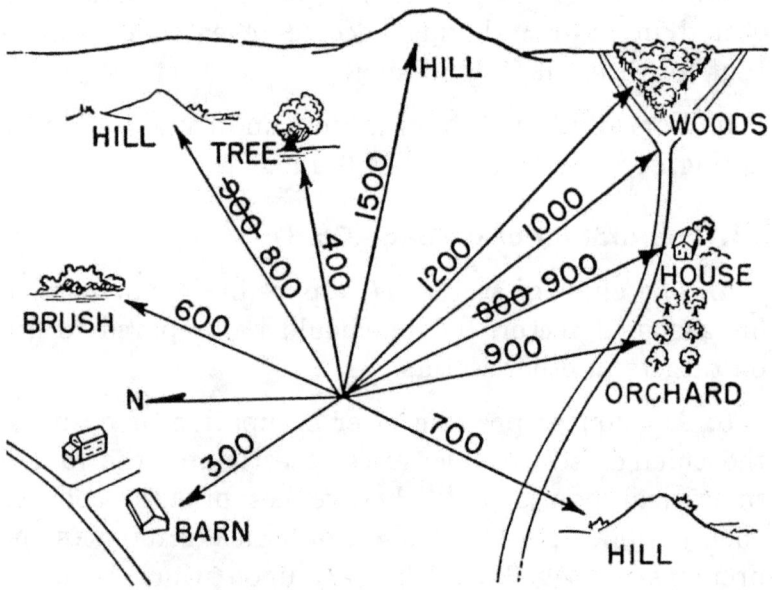

Figure 12. Range card for direct laying.

personnel, whether alone or accompanying tanks, will seldom present themselves as a clearly defined target. Normally, attacking troops, using all available cover, reveal themselves only fleetingly. Accordingly, fire is conducted on the area containing the attackers rather than on the individuals. Tanks usually attack in groups and may be accompanied by infantry. Normally, first priority is given to attack of those targets within the assigned sector of the weapon and second priority to targets in other sectors. Priority within the assigned zone is given to—

a. Tanks at short ranges, threatening to overrun the position.

b. Hull down stationary tanks covering the advance of other tanks.

c. The commander's tank, if identified.

d. The tank nearest to cover which may disappear and reappear at unexpected places.

e. The rear tank of a column moving across the front of the position, to minimize the possibility of attracting attention of the tank column to the battery position.

100. Ammunition

a. General. For close-in fires, a variety of fuzes and shells are available (TM 9–1901). In using AP or HE shell, the maximum charge is used habitually for speed, ease in adjustment, imparting forward motion to fragments, and more effective fuze action. The flat trajectory resulting from use of the maximum charge, coupled with dug-in weapons, may make extremely close-in fire difficult owing to projectiles skipping without detonating on impact. In certain areas throughout the sector of fire, fuzes may fail to function because the slope of the ground and the angle of fall of the shell are nearly parallel and the fuze may not strike the ground. Location of these areas cannot be predicted, since they depend upon the characteristics of the ammunition fired, the range, the slope and hardness of the ground, and the height of the tube above the ground. When this condition is known or suspected, preparation of sectors of fire may remedy the situation. The terrain may be prepared for direct fire by placing mounds of sandbags, dirt, or logs in the sector of fire. When direct fire is placed on these or other previously selected points as they are approached by an attacking force, the probability of obtaining effective bursts is increased.

b. Projectiles. Projectiles for the 155-mm gun may be armor-piercing (AP); high explosive (HE); or smoke, white phosphorus (WP). AP will be used against armored vehicles and like targets. Shell HE is used against troops, unarmored targets and, when AP is not available, armored targets. WP may be used to set immobile tanks and vehicles on fire, to restrict defiles, and to cause casualties. However, personnel firing WP must keep in mind the effect of the resulting smoke screen on friendly elements. The only projectile for the 8-inch howitzer is shell HE, and it is used against all types of targets.

c. Fuzes. Shell HE may be fuzed with fuzes quick, delay, VT time, or concrete-piercing. Fuzes for AP ammunition are inclosed in the base of the shell and cannot be set. Fuzes for WP ammunition are point detonating (PD), M51-series, and may be set before firing for superquick or delay action.

    (1) Fuze quick is the best selection for use with HE shell on close-in targets. It is highly effective and, since no fuze setting is required, is much faster to use.

    (2) The time required to set the fuze and to adjust the point of impact for maximum ricochet effect against personnel makes fuze delay less desirable than fuze quick for close-in targets. Because of its penetrating ability, fuze delay is effective in direct fire against prepared or fixed positions or against immobile tanks.

    (3) Fuze time is the least desirable fuze for close-in targets. At short fuze settings, variations in time of burning give wide range dispersion of bursts. Hence, this fuze should be used only for ranges of more than 1,000 yards. The areas covered effectively by air and ricochet bursts are similar.

    (4) Concrete-piercing fuze with shell HE should be used against concrete pillboxes or fortifications.

    (5) Fuze VT is not ordinarily used for close-in fires.

101. Trajectories

Trajectory characteristics change with the type of ammunition and the charge fired. The following trajectory characteristics govern the conduct of fire by direct laying when the highest charge is used for these weapons.

a. Ranges from 0 to 600 yards (155-mm gun) using supercharge. Ranges from 0 to 400 yards (8-inch howitzer) using charge 7. Within these range limits the trajectory will be flat enough to prevent an 8-foot tank from passing safely under it. Fields of fire and terrain allowing, the upper range limit is the ideal at which to open fire by direct laying. Direct fire can then be conducted over the maximum time without misses if deflection is correct. After hits are obtained, the point of burst should be moved to a vital part of the tank. Also, there is less risk of screening the target with the smoke from a short burst.

b. Ranges from 600 to 2,500 yards (155-mm gun) using supercharge. Ranges from 400 to 1,400 yards (8-inch howitzer) using charge 7. These range limits include the zone in which a reasonably short adjustment can be conducted. Range changes of 200 yards should be made to establish a bracket if the target is near a reference point of known range. Range changes of 400 yards should be made if ranges to reference points have been estimated or if the range to the target is estimated. Fire should not be opened at these ranges unless surprise is not important.

c. Ranges from 1,500 to 2,500 yards (8-inch howitzer) using charge 7. This zone includes the ranges at which hits by direct fire are only reasonably possible. Bracket methods are normally used to obtain adjustment in this zone. There is more dispersion in this zone; direct fire should not be opened at these ranges unless surprise is not important. Indirect fire will usually give better results on stationary targets or on slow moving targets such as infantry.

d. Ranges over 3,000 yards (155-mm gun) using supercharge. Ranges over 2,500 yards (8-inch howitzer) using charge 7. At these ranges firing by direct laying is not advisable against moving targets. Dispersion is the controlling factor. Ranges must be known accurately or determined by bracketing. At ranges over 3,000 yards, the slope of fall of the projectile becomes so great that a hit on a moving target is very difficult to obtain.

102. Vertical Displacement Table

Vertical displacement is the change in the point of burst (up or down) between two rounds fired with different ranges at an upright target. The vertical displacement for a 100-yard range change at various ranges is shown in table IV (155-mm gun, shell HE, M101, supercharge) and in table V (8-inch howitzer, shell HE, charge 7).

Table IV. Vertical Displacement (Feet) per 100-Yard Range Change (155-mm Gun)

Range (yards)	Displacement, feet, shell HE, M101, supercharge	Remarks	Range (yards)
100	0	Start firing, using 600-yard range setting.	100
200	0.5		200
300	0.5		300
400	1.0		400
500	1.0		500
600	1.0	Start firing, using estimated range. Increase or decrease by 50- or 100-yard changes. Bracketing not necessary.	600
700	1.5		700
800	1.5		800
900	2.0		900
1,000	2.0		1,000
1,100	2.0		1,100

Table IV. Vertical Displacement (Feet) per 100-Yard Range Change (155-mm Gun)—continued

Range (yards)	Displacement, feet, shell HE, M101, super-charge	Remarks	Range (yards)
1,200	2.5	Start firing, using estimated range. Increase or decrease by 50- or 100-yard changes. Bracketing not necessary.	1,200
1,300	2.5		1,300
1,400	3.0		1,400
1,500	3.0		1,500
1,600	3.0		1,600
1,700	3.5		1,700
1,800	3.5		1,800
1,900	3.5		1,900
2,000	4.0		2,000
2,100	4.5		2,100
2,200	4.5		2,200
2,300	4.5		2,300
2,400	5.0		2,400
2,500	5.3		2,500
2,600	5.6		2,600
2,700	5.8		2,700
2,800	6.1		2,800
2,900	6.2		2,900
3,000	6.5		3,000
Over 3,000		At ranges over 3,000 yards, direct laying is too inaccurate to be used against moving targets.	

Table V. Vertical Displacement (Feet) per 100-Yard Range Change (8-inch Howitzer)

Range (yards)	Displacement, feet, shell HE, charge 7	Remarks	Range (yards)
100	0.5	Start firing, using 400-yard range setting.	100
200	1.0		200
300	1.5		300

Table V.—continued

Range (yards)	Displacement, feet, shell HE, charge 7	Remarks	Range (yards)
400	2.0	Start firing, using estimated range. Increase or decrease by 50- or 100-yard changes. Bracketing not necessary.	400
500	2.5		500
600	3.0		600
700	3.5		700
800	4.0		800
900	4.0		900
1,000	4.5		1,000
1,100	5.0		1,100
1,200	5.5		1,200
1,300	6.0		1,300
1,400	6.5		1,400
1,500	7.0		1,500
1,600	7.5	Bracket target (get bursts over and short) to obtain hit.	1,600
1,700	8.0		1,700
1,800	8.5		1,800
1,900	9.0		1,900
2,000	9.5		2,000
2,100	10.0		2,100
2,200	10.5		2,200
2,300	11.5		2,300
2,400	12.0		2,400
2,500	12.5		2,500
Over 2,500		At ranges over 2,500 yards, direct laying is too inaccurate to be used against moving targets.	

Section II. DUTIES OF CHIEF OF SECTION

103. List of Duties

(Detailed description of duties, par. 104–109.)
a. Conducts the fire of his piece.

   b. Identifies or selects the target.

   c. Estimates the range to the target.

   d. Determines the lead in mils.

   e. Gives initial commands.

   f. Gives subsequent commands based on observed effect.

104. Conducts the Fire of His Piece

The chief of section conducts the fire of his piece when the executive commands TARGET (SO-AND-SO), FIRE AT WILL, or simply FIRE AT WILL.

105. Identifies or Selects the Target

If the executive designates an object or one of a group of objects as the target, the chief of section must correctly identify this target. If the target is a group of tanks or other objects, the chief of section selects the one that, in his estimation, is the greatest threat to his own position or the position of the supported troops. He repeats the identification to his section, using the minimum number of words such as LEAD TANK or MOVING TANK.

106. Estimates the Range to the Target

Range cards (fig. 12) with accurately measured ranges to key points provide the best means for determining the initial range. If a range card has not been prepared, the range is estimated.

107. Determines the Lead in Mils

The appropriate lead in mils for targets moving at various speeds for firing with maximum charge is as follows:

Lateral speed	Lead (mils)	
	155-mm gun	8-inch howitzer
Under 10 MPH	5	5
10 MPH or over	5	10

108. Gives Initial Commands

The chief of section gives fire commands containing the following elements in sequence:

a. Designation of Target. The command is TARGET (SO-AND-SO). Identification must be clear and unmistakable and should employ the minimum number of words.

b. Projectile, Charge, and Fuze. The chief of section selects the appropriate projectile, charge, and fuze and commands SHELL (SUCH-AND-SUCH), CHARGE (SUCH-AND-SUCH), if applicable, and FUZE (SUCH-AND-SUCH).

c. Lead. The command is LEAD (SO MUCH). See paragraph 107 for the method of determining lead.

d. Method of Fire. Fire is continuous unless otherwise commanded. In continuous fire, the piece is loaded and laid as rapidly as possible and fired at the command of the gunner.

e. Range. The command is RANGE (SO MUCH). The range commanded by the chief of section is that range to be set on the sight reticle. For determining range, see paragraph 106.

Note. (155-mm gun only.) When the chief of section commands RANGE (SO MUCH), he must consider the ammunition being used. The elbow telescope T159 is graduated for shell HE, M101, supercharge.

109. Gives Subsequent Commands Based on Observed Effect

a. The chief of section gives the following commands based on observed effect:

(1) *Change in lead.* During adjustment, the lead in mils is changed by the command RIGHT (LEFT) (SO MUCH).

(2) *Change in range.* During adjustment, the range is increased by the command ADD (SO MUCH) and decreased by the command DROP (SO MUCH). See paragraph 102 for the method of determining changes in range during adjustment of fire.

b. When the breechblock is closed, the chief of section gives further changes in firing data based on movement of the target during the time required for loading.

Section III. DUTIES OF REMAINDER OF SECTION

110. Gunner, List of Duties

(Detailed description of duties, par. 111–116.)

a. Levels cant correction level vial.

b. Lays for direction and elevation on a stationary target.

c. Takes the proper lead and range on a moving target.

d. Fires the piece.

e. Continues tracking the target.

f. Applies corrections for lead and range.

111. Levels Cant Correction Level Vial

The gunner levels the cant correction level vial by rotating the can correction knob. This brings the telescope reticle pattern into plumb, correcting for cant when the motor carriage is parked on a slope.

112. Lays for Direction and Elevation on a Stationary Target

When firing directly at an immobile target, the gunner, sighting through the direct laying telescope, traverses the tube until the vertical line in the telescope reticle appears on the center of the target (at close ranges, on a port, slit, or some other vulnerable point of the target).

113. Takes the Proper Lead

When the chief of section commands LEAD (SO MUCH), the gunner, sighting through the direct laying telescope, traverses the piece so that the target appears in the reticle traveling toward the vertical center line at the correct range (fig. 13). The lead expressed in mils determines the distance between the center of the target and the vertical center line. The gunner maintains this lead continuously until the target is immobilized or the chief of section announces a change. Tracking is best accomplished with low speed power if the mechanism is in smooth operating condition. If tracking is jerky, manual traverse should be used.

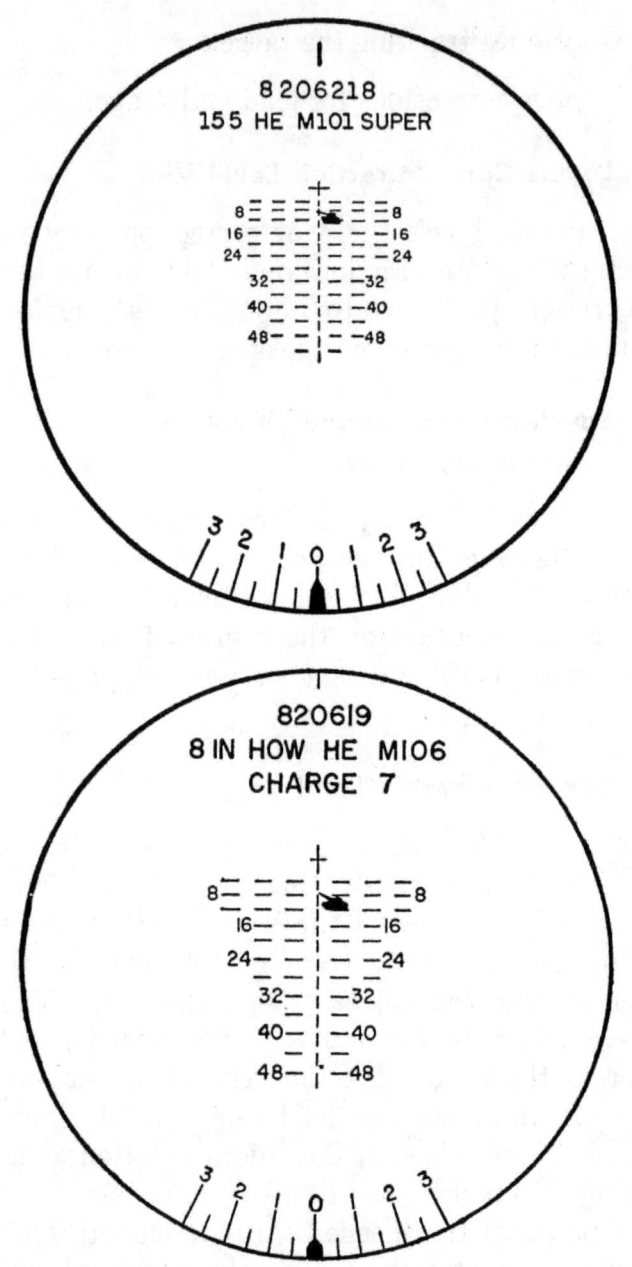

Figure 13. Gunner's sight picture, direct laying telescope, lead 5 mils, range 1,200 yards.

114. Fires the Piece

To fire the piece the gunner takes those actions prescribed in paragraph 47.

115. Continues Tracking the Target

The gunner continues to track the target until CEASE FIRE is given.

116. Applies Corrections for Lead

The gunner applies the chief of section's command for changing the lead and range (par. 109).

117. Remainder of Section

a. Cannoneers. Except as indicated in *b* below, the remainder of the section perform duties as prescribed for indirect laying (par. 21–95).

b. Driver, Motor Carriage. The driver takes his post in the driver's compartment, starts the engine, and prepares to shift the vehicle as directed.

CHAPTER 7

MISCELLANEOUS PROCEDURES AND TECHNIQUES

118. Precision in Laying

a. Sighting and laying instruments, fuze setters, and elevating and traversing mechanism must be properly operated to reduce the effects of lost motion. For uniformity and accuracy, the last motion in setting instruments must be from lower to higher numbers; final motion of the elevating handwheel must be in the direction of more difficult movement, and final motion in traversing must be from left to right. Personnel who lay the piece must be required to verify the laying after the breech has been closed.

b. When a scale is set and read on the panoramic telescope or a bubble is centered, the line of sight must be at a right angle to the scale or level vial to prevent parallax errors. Bubbles must be centered exactly.

c. For uniformity and accuracy in laying on aiming posts, the vertical cross hair in the reticle of the panoramic telescope must be alined with the left edges of the aiming posts.

119. Aiming Points

a. General. After the piece has been laid initially for direction, it is referred to the aiming posts and usually to one or more distant aiming points as described in paragraph 42. An aiming point must have a sharply defined point or vertical line which is clearly visible from the piece so that the cross hairs of the panoramic telescope can be alined on exactly the same place each time the piece is relaid.

b. Distant Aiming Point. A distant aiming point is one at sufficient distance so that normal displacements of the piece in firing or traverse will not cause a horizontal angular change in direction (with the same settings on the azimuth scales) of more than one-half mil. This distance should be at least 3,000 yards. The executive officer usually designates the distant aiming point or points to be used.

120. Aiming Posts

a. Two aiming posts are used for each piece. Each aiming post is equipped with a light for use at night. The most desirable distance from the piece to the far aiming post is 100 yards, considering accuracy of laying, visibility, and ability to control the aiming post lights. First, the far aiming post is set up and alined. The near aiming post is then set up at the midpoint between the far aiming post and the piece and is alined by the gunner so that the vertical cross hair of the telescope and the left edge of the two aiming posts are in alinement. To insure equal spacing of aiming posts, the distance to both the near and far aiming post should be paced by the same man. If ground conditions make pacing inaccurate, the distance from the piece to the aiming posts may be measured by using the panoramic telescope, with the aiming posts as measuring devices (*d* below).

b. For night use, the aiming post lights should be adjusted so that the far light will appear several feet above the near light. On flat terrain this may be accomplished by using only the lower half of the near aiming post. The two lights placed in this way will establish a vertical line for laying the piece.

c. Since the panoramic telescope is mounted at considerable distance from the center of rotation of the top carriage, large changes in deflection will cause misalinement of the aiming posts. Placing the aiming posts to the right front at a deflection of 600 when the piece is in the center of traverse will keep this misalinement to a minimum and still allow for maximum visibility.

d. To measure the distance from piece to aiming posts, the stadia method may be employed by using the panoramic telescope and the aiming post as measuring devices. No. 5 cannoneer, in setting out the aiming posts, holds the upper section of one of the aiming posts in a horizontal position, perpendicular to the line of sighting. The gunner measures the length of this section in mils using the reticle of the panoramic telescope. For example, the upper section of the aiming post is 4½ feet long and measures 15 mils when it is 100 yards from the piece. The proper location for the near aiming post, in this case, would be at the point at which the 4½-foot section measures 30 mils. In many cases, the ideal spacing of 50 and 100 yards cannot be obtained, but the aiming posts will be properly spaced when the near aiming post is set at a point where the 4½-foot section measures twice the number of mils it measured at the far aiming post location. This measurement may be performed at night by attaching the night lighting devices at the 4½-foot marks on the aiming posts.

121. Correction for Displacement of Aiming Posts

When the gunner notes that the vertical line of the telescope is displaced from the line formed by the two aiming posts (or aiming post lights), he lays the piece in such a manner that the far aiming post (light) appears exactly midway between the near aiming post (light) and the vertical cross hair (fig. 14). If the displacement is due to traversing the piece, the gunner continues to lay as described above. If the displacement is due to progressive shifting of the carriage from shock of firing or other cause, the gunner will notify the chief of section, who, at the first lull in firing, will notify the executive and request permission to realine the aiming posts. To realine the aiming posts, the piece is laid with the far aiming post midway between the near aiming post and the vertical cross hair (fig. 14). The far aiming post is moved into alinement with the vertical cross hair of the telescope and then the near aiming post is alined. If terrain conditions make it impracticable to move 1 of the 2 aiming posts, the piece is laid for direction and referred to the aiming post that cannot be moved. This deflection is reported to the executive. The other post is alined by using the method described in paragraph 120, and the azimuth counter is turned to retain the same deflection that was used prior to realinement of the aiming posts.

122. Testing Targets

Testing targets will be more useful if the following improvements are made:

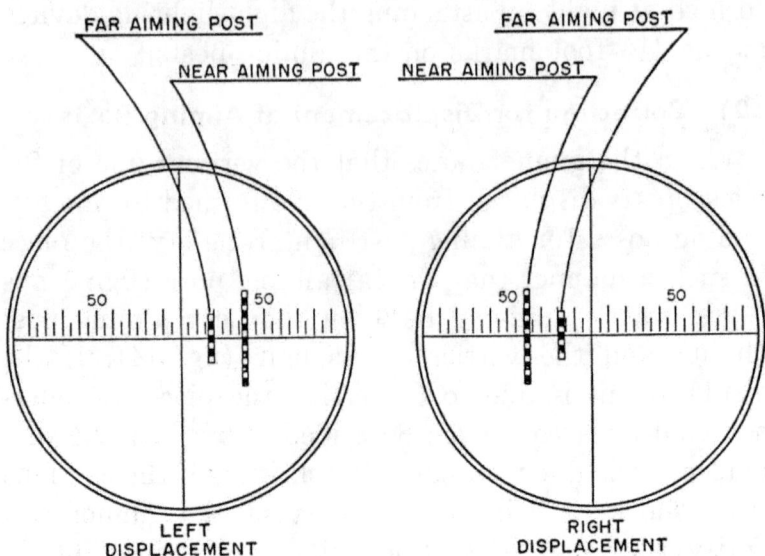

Figure 14. Gunner's sight picture of aiming posts in proper relationship when correcting for displacement.

a. The target should be mounted on a flat piece of masonite, wallboard, or similar material.

b. To insure stability of the testing target throughout bore sighting, it should be fastened securely to a stand.

c. For use in either leveling or canting the testing target, a mil scale may be inscribed at the bottom of the target. A small nail at the top marks the center from which the arc was drawn and provides a hook from which to suspend the plumb line.

d. Vertical reference lines (fig. 15) may be drawn through the centers of each of the diagrams. These lines may be used when the trunnions cannot be leveled by setting the testing target with the cant angle of the piece. The target is tilted until the line of sight through the tube tracks between the tube reference line when the tube is elevated or depressed. Then, the panoramic telescope should be adjusted so that its vertical cross hair tracks between the appropriate reference lines when the tube is elevated or depressed.

e. To facilitate bore sighting in darkness, a $\frac{1}{16}$-inch hole may be bored through the mounted testing target at the center of each aiming diagram. A flashlight held against the target behind the appropriate hole pro-

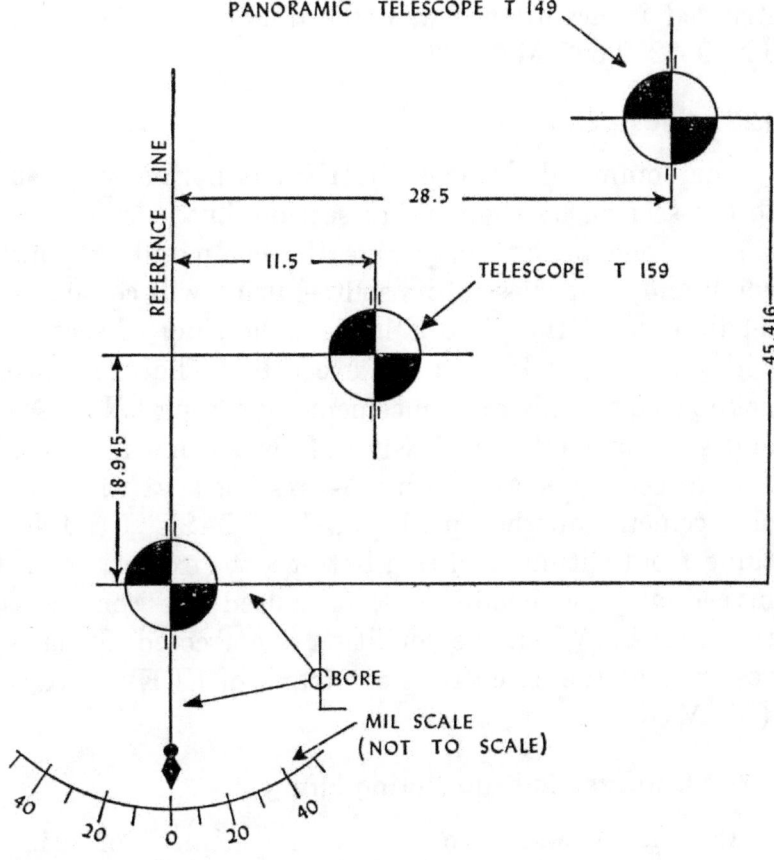

Figure 15. Vertical reference lines.

TAGO 7231C—July 75

vides an aiming point for use in blackout conditions. Fasten patches of felt padding on the back of the target covering the regions of each hole so that light from the flashlight will not escape. The flashlight must be lit only after it is placed firmly in position. Care must be taken to prevent disturbing the position of the testing target.

f. If the proper testing target is not available, a substitute with aiming diagrams for the bore, subcaliber piece, and panoramic telescope may be constructed in accordance with the dimensions shown in TM 9–7212 or TM 9–7220.

123. Cease Firing

The command CEASE FIRING is normally given to the section by the chief of section, but in emergencies anyone present may give the command. At this command, regardless of its source, firing will cease immediately. If the piece is loaded, the chief of section will report that fact to the executive. The executive acknowledges this announcement by saying "No. (so-and-so) loaded." If CEASE FIRING came from the fire direction center, firing is resumed at the announcement of the quadrant. If CEASE FIRING came from within the firing battery, the executive will investigate the condition that caused the command to be given. When the condition is corrected, firing is resumed at the executive's command of ELEVATION (SO MUCH).

124. Changes in Data During Firing

If it is necessary to correct any element of firing data, all firing previously ordered but not yet executed is stopped by the command CEASE FIRING. Corrected data is then announced. If the piece is not loaded, the new data will be set off and firing resumed at the command ELEVATION (SO MUCH). If the piece is loaded and no change in fuze setting is required or if the piece is loaded with percussion-fuzed shell, the new data is set off and firing is resumed at the command ELEVATION (SO MUCH). If the piece is loaded with time-fuzed shell and the data requires a change in fuze setting, the chief of section will suspend firing and that fact will be reported to the executive; for example, "No. 2 loaded, time (so much)." The piece will not be unloaded unless so directed by the executive. In continuous fire, changes in data are so applied as not to stop the fire or break its continuity.

125. To Unload the Piece

A complete round, once loaded, should always be fired in preference to being unloaded unless military necessity dictates otherwise. The piece will be unloaded only upon the specific orders and under the direct supervision of an officer. To unload the 155-mm gun (par. 91, TM 9-7212) and the 8-inch howitzer (par. 91, TM 9-7220), the command is UNLOAD, and the operation is performed as follows:

a. The chief of section has the primer and firing mechanism removed, the breech opened, the powder charge withdrawn, the chamber filled with waste, the breech closed, and the tube depressed to zero elevation.

b. No. 4 and 5 insert the unloading rammer into the muzzle end of the tube and push carefully until the rammer head encircles the fuze and contacts the projectile. Steadily increasing pressure is applied by tapping the end of the rammer staff with a wooden block, if necessary, to loosen the projectile.

c. When the projectile is loosened, No. 4 and 5 suspend operation of the rammer while the chief of section has the breech opened and the waste removed. No. 3 then places loading trough in position in the breech to receive the projectile as the tube is elevated to the load position, approximately 175 mils.

d. The chief of section, standing at the breech end of the bore, holds a section of rammer staff, if available, or a similar item, firmly against the base of the projectile. He steadies its backward movement as No. 1 and 2 push the projectile onto the loading trough.

e. After it is unloaded, the ammunition specialist disposes of the projectile as directed by the chief of section.

f. For further information on unloading, see FM 6–140, TM 9–7212, and TM 9–7220. For instruction concerning misfires, see paragraphs 90 and 180, TM 9–7212 and TM 9–7220.

126. Care of Ammunition

a. To insure uniform results in firing, to prolong the life of the tube, and to avoid accidents, care must be exercised in the storage and handling of ammunition at the battery. Provisions of TM 9–1900 applicable to field service should be followed carefully. In the field, conditions existing in each position will determine the amount of time, labor, and materials required to store and preserve the ammunition adequately. If the position is to be occupied for only a few hours, a tarpaulin spread on the ground may be sufficient; for longer periods of time, more elaborate facilities should be provided.

b. Ammunition must be protected from damage. When projectiles are received, they should be sorted into lots and placed in the best available storage. Ammunition data cards should be kept until after all ammunition for that lot is expended. The eyebolt lifting plug should not be removed from unfuzed projectiles until the fuze is to be inserted. Protection should be provided against moisture, dirt, direct rays of sun and, as far as practicable, artillery fire and bombing. Protection against weather, dirt, and sun may be obtained by the use of tarpaulin and dunnage. Projectiles stacked in the open should be raised off the ground at least 6 inches. If drainage is not good, ditches should be dug around the stacks. A liberal use of dunnage should be made between layers, and covering tarpaulins should be raised at least 6 inches from the stack to insure adequate ventilation. Ammunition for the 155-mm gun should be placed in stacks not more than 3 layers high and contain not more than 50 rounds; ammunition for the 8-inch howitzer should have only a 1-layer stack and contain not more than 25 rounds. Stacks should be at least 10 yards apart.

c. Powder charges should be sorted into lots and protected from sources of high temperatures, including direct rays of the sun. More uniform firing is obtained if the charges are of the same temperature. Powder charges should not be removed from containers until just before firing.

d. Explosive elements in primers and fuzes are particularly sensitive to shock and high temperature; therefore, strict attention should be given to their care and handling. Protection and safety devices should not be removed from fuzes until just before use. No attempt should ever be made to disassemble a fuze into its components.

e. For further information on care of ammunition, see FM 6-40, FM 6-140, TM 9-1900, TM 9-1901, TM 9-7212, and TM 9-7220.

127. Section Data Board

When positions are occupied for more than a few hours, data boards may be used by each section for recording such items as deflections to aiming points, calibration corrections when appropriate, minimum elevations, data for the barrage and counter-preparations, and other data which may be needed quickly. If such information assumes a standard pattern, the section may paint a form on the inside of the hull or other convenient part of the weapon and chalk in the various items of information in the appropriate spaces.

CHAPTER 8

BORE SIGHTING

Section I. GENERAL

128. Description

Bore sighting is the process of verifying that the optical axis of the on-carriage fire control equipment is parallel with the axis of the tube of the weapon, both for deflection and for elevation. Any misalinement discovered through bore sighting is corrected as described in paragraphs 133 through 137. The tube should be placed near its center of traverse prior to bore sighting. All instruments and mounts must be positioned securely; there must be no free play. Bore sighting is conducted before firing and, when necessary, during lulls in firing.

129. Equipment

The following equipment is needed for bore sighting:

a. Bore Sights. Front and rear bore sights or improvised substitutes are necessary for all but the standard angle method for bore sighting. If bore sights are not available, cross hairs may be fastened on the muzzle, and the obturator spindle vent may be used as a rear sighting guide.

b. Testing Target. A testing target or suitable substitute is needed for preparatory steps in testing and for certain methods of bore sighting. If a testing target is not available, a clearly defined aiming point 3,000 or more yards from the piece may be used to accomplish approximately the same purpose as the testing target.

c. Tools. The section equipment includes all the necessary tools for bore sighting and testing. If any item of sighting and fire control equipment fails to meet the prescribed tests, ordnance maintenance personnel must be notified.

130. Conditions

The on-carriage fire control equipment is in correct alinement when the conditions in *a* through *d* below exist.

a. Mounts and instruments are securely attached, and there is no binding or excessive backlash between the gears.

b. The lines of sight of on-carriage sighting equipment are parallel to the axis of the bore throughout the limits of elevation.

c. All scales and indexes read zero.

d. All bubbles are leveled.

131. Leveling

a. Trunnions. Although it is not absolutely necessary to level the trunnions for bore sighting, it is advisable to do so whenever possible. Accurate results can be obtained more readily if the trunnions are level, because then a tilt corresponding to the cant does not have to be introduced in the telescope mount and the testing target when used. The trunnions can be leveled by moving the carriage to level ground or by building up the standing for one of the motor carriage tracks. In no case should there be more than 20 mils cant.

b. *Plumb Line.* The best method to check leveling is by means of the plumb line. The line is suspended directly in front of the axis of the bore at a distance of 15 feet. The line of sight should track the plumb line as the tube is depressed and elevated between minimum elevation and the limits described by a plumb line which is as long as practicable. The plumb line must be shielded from wind currents, and the plumb bob or weight should be suspended in a container of liquid in order to keep the long plumb line taut.

c. *Gunner's Quadrant.* In leveling operations in which the gunner's quadrant is used, a quadrant that has been tested (par. 147–149) and found to be accurate is required.

132. Methods

There are three methods for bore sighting these weapons.

a. Testing target method (par. 133–137).

b. Distant aiming point method (par. 138 and 139).

c. Standard angle method (par. 140–143).

Section II. TESTING TARGET METHOD

133. General

The testing target method of bore sighting consists of using the aiming diagrams of the testing target as aiming points. The preliminary steps in bore sighting are as follows:

a. Trunnions. Level the trunnions as described in paragraph 131.

b. Tube. Level the tube by using the gunner's quadrant on the leveling plates of the breech ring. Make certain that the shoes on the gunner's quadrant are positioned between the engraved lines on the leveling plates.

c. Bore Sights. Open breech and insert breech bore sight in the chamber. Attach the muzzle bore sight, stretching linen cords across the witness marks and over the cords on the muzzle and securing the ends by placing a strap around the end of the muzzle. If breech bore sight is not available, the obturator spindle vent may be used.

d. Testing Target Alinement. The testing target normally should be located at least 50 yards in front of the muzzle. If the trunnions are level, level the testing target by means of a plumb line or the vertical reference lines. If the trunnions are not level, cant the target to correspond to the cant of the trunnions. In either case, the face of the target is perpendicular to the axis of the bore, both laterally and longitudinally. Without moving the piece, except for elevating and depressing slightly when using testing target reference lines, aline the tube testing target diagram with the line of sight through the tube.

e. Telescope Mount T179E2. The level of the locating plate is checked by insuring that the bubbles in the leveling segment level vials and the leveling mechanism level vials are accurately centered.

134. Elevation Counter

To make the elevation counter reading agree with

the tube elevation, place the gunner's quadrant on the leveling plates and, by elevating or depressing, bring the tube to *0* elevation. Rotate the elevation counter setting knob until the elevation vernier indexes are alined. If the counter reading does not agree with the gunner's quadrant reading within plus or minus ½ mil, turn the elevation counter setting knob to 0, loosen the lock setscrew on the side of the bellcrank (fig. 16) and, with a $5/32$-inch hex setscrew wrench, turn the bellcrank adjusting screw until the elevation vernier indexes are alined. Tighten the lock setscrew, making certain the adjustment does not change.

Note. Some T179 telescope mounts will have a setscrew or cap screw on top of the bellcrank adjusting screw. This screw must be removed before the bellcrank adjustment can be made.

135. Panoramic Telescope Alinement

Rotate the elevation and azimuth knobs until the horizontal and vertical lines of the reticle pattern are alined with the horizontal and vertical lines respectively on the panoramic telescope diagram on the testing target. The coarse azimuth and micrometer scales should now read 0 within 5 mils; if they do not, loosen the 3 setscrews on the micrometer scale and slip the scale to 0 with the cross hairs of the telescope on the center of the telescope diagram. If the coarse azimuth and micrometer scales should be in error by more than 5 mils, the telescope being properly seated, adjustment must be made by ordnance maintenance personnel.

136. Direct Fire Telescope Alinement

After procedures prescribed in paragraph 135 are

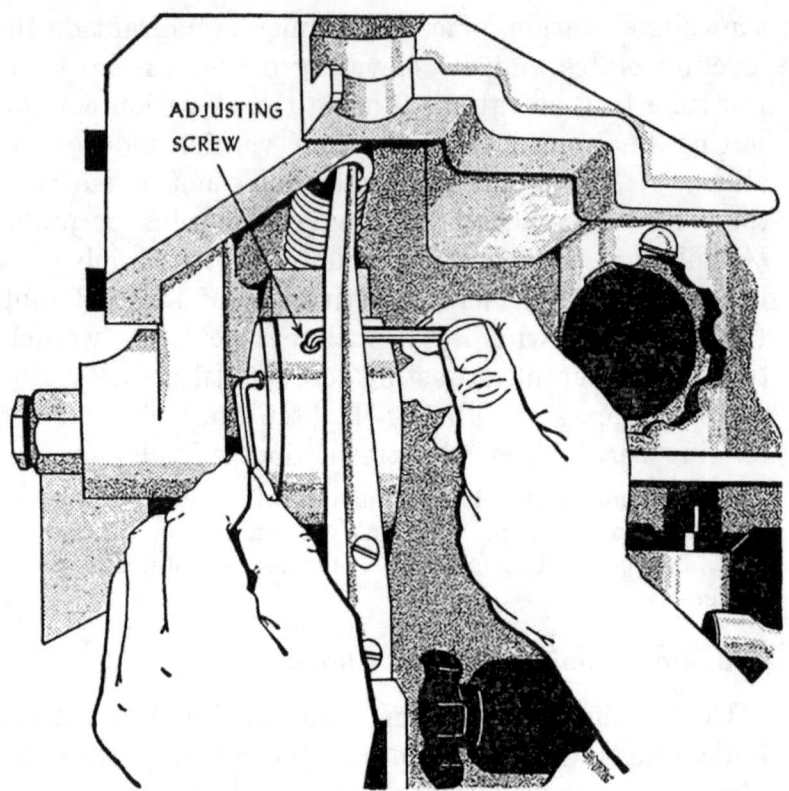

Figure 16. Bellcrank adjustment.

performed, rotate the cant correction knob to level the telescope by centering the bubble in the cant correction vial. Unlock the bore sight knob locking levers. Rotate the azimuth and elevation bore sight knobs until the cross hairs at zero range on the rectile pattern are alined with the center of the direct fire telescope diagram on the testing target. In the event that the cross hairs of the telescope cannot be brought into coincidence with the lines on the aiming diagram, within limits of the bore sight knobs, notify ordnance maintenance personnel.

137. Periscope M15A1 Alinement

For information on procedures to be followed in bore sighting the periscope M15A1, see TM 9-7212 or TM 9-7220.

Section III. DISTANT AIMING POINT METHOD

138. General

The distant aiming point method consists of alining the optical axis of the on-carriage fire control equipment and the line of sight through the tube on a common point at least 3,000 yards from the piece and as near 0 elevation as possible.

139. Procedure, Distant Aiming Point Method

The steps prescribed for the testing target method apply to the distant aiming point method, except that the bore sights and optical sights are alined on the same point instead of on the diagrams of the testing target. Accurate cross-leveling of the trunnions is unnecessary when bore sighting on a distant aiming point, because the lines of sight converge on a single point.

Section IV. STANDARD ANGLE METHOD

140. General

When conditions exist to make other methods of bore sighting impracticable, the standard angle method may be used. In this method, the alinement of the optical axis of the panoramic telescope parallel to the axis of the bore is tested and adjusted by referring to a selected point on the muzzle. The deflection and elevation angles necessary to refer the line of sight of the telescope to the selected point on the muzzle are referred to as the standard angles. Once standard angles have been determined, they may be used for a quick test of the alinement of the panoramic telescope when more precise methods cannot be used. Correction of misalinement, as a result of this test, should be verified by a more accurate method at the earliest opportunity. When the standard angle method of bore sighting is being used, the position of the recoiling parts with respect to the nonrecoiling parts must be the same as when the standard angles were determined. Therefore, the recoil mechanism must be checked to see that it contains the proper amount of recoil oil before determining the standard angles. Standard angles are usable only as long as the same tube-carriage combination is intact. If interchange of tubes or carriages is made, standard angles must be reestablished.

141. Parallax

Parallax in the panoramic telescope must be eliminated. This is done by placing in front of the eyepiece lens a dark colored cardboard or metal parallax shield of the same diameter as the eyepiece lens housing. The shield should have an exactly centered hole $\frac{1}{16}$-inch in diameter. A more permanent parallax shield may be constructed of brass or bronze shim stock. When the shield is constructed of metal, a series of fingers approximately $\frac{3}{16}$-inch wide and $\frac{1}{4}$-inch long separated by $\frac{1}{4}$-inch spaces should extend beyond the perimeter of the shield. These fingers should be bent along the circumference of the circle until they form an angle of 90° with the surface of the shield. The fingers serve as a means of clipping the shield in place quickly and permit easy removal. If the eyepiece has a rubber eyeguard, the fingers permit alinement within the guard without its removal.

142. Preliminary Operations

The ideal time to determine the standard angles for later use is after performing basic periodic tests when the trunnions are level and the panoramic telescope mount is known to be in correct alinement. Procedure for determining standard angles is as follows:

a. With the tube in battery, scribe lines in the paint to mark the normal position of parts which move in recoil with respect to parts which do not move in recoil.

b. Bore sight the piece by using a testing target or distant aiming point.

c. With friction tape, fasten a bright common pin in the right horizontal witness mark. Allow the pin to project to the right of the muzzle (fig. 17).

d. Fasten the parallax shield over the eyepiece of the panoramic telescope.

e. Verify that the 0 elevation indicator on the telescope is at 0 and level the telescope mount.

f. By turning the azimuth micrometer knob and elevating or depressing the tube as necessary, place the cross hairs of the sight on the metal pin in the right horizontal witness mark of the tube.

g. Verify that the telescope mount is level and that the horizontal and vertical lines of the telescope are exactly on the junction of the pin with the muzzle.

h. Read and record the deflection from the panoramic telescope to the nearest ¼ mil. (Since the

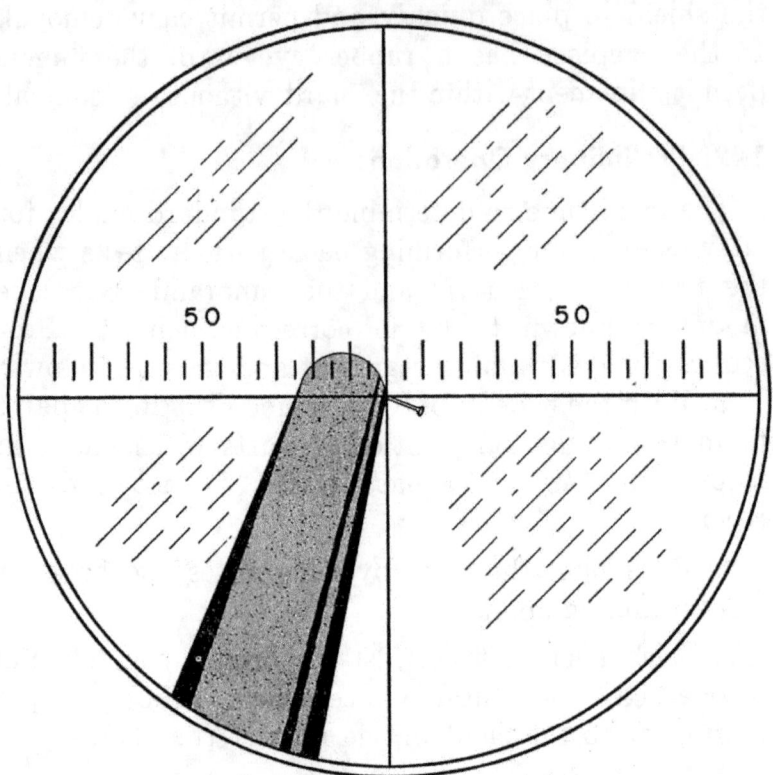

Figure 17. Sight picture of projecting pin.

graduations are to the nearest mil, it is necessary to interpolate to the nearest ¼ mil.) This is the standard azimuth angle for the piece tested.

i. With the gunner's quadrant seated on the quadrant seats, measure and record the elevation of the tube to the nearest ¼ mil. This is the standard elevation angle for the piece tested.

143. Procedure

Once the standard angles have been determined and recorded, the steps in performing the standard angle method of bore sighting are as follows (*a–f* below):

a. Verify that the parts that move in recoil are in the same position with respect to the nonrecoiling parts as they were when the standard angles were determined. If they are not in the same position, the amount of recoil oil in the recoil mechanism must be modified until the distance from the end of the breech to the scribed line is the same as in paragraph 142*a*.

b. With friction tape, fasten a bright common pin in the right horizontal witness mark so that the pin projects out to the right of the muzzle.

c. Place the parallax shield on the eyepiece of the telescope.

d. Set off the standard elevation angle (par. 142*i*).

e. Set off the standard azimuth angle on the panoramic telescope (par. 142*h*).

f. If the intersection of the cross hairs of the panoramic telescope is not exactly on the junction of the pin and the muzzle, the sight is out of adjustment. If the azimuth angle is in error, it may be corrected by section personnel by slipping the azimuth micrometer scale. If the elevation angle is in error by more than $\frac{1}{2}$ mil, the bellcrank adjustment (par. 134) must be performed.

CHAPTER 9

BASIC PERIODIC TESTS

Section I. GENERAL

144. Purpose and Scope

a. The purpose of this chapter is to describe the procedures for making basic periodic tests of on-carriage fire control equipment. The procedures covered include only those that may be accomplished at battery level. It is not contemplated that using units will have the necessary facilities, tools, or skilled personnel to perform the more precise tests and adjustments of sighting and fire control equipment. If the elevation counter, telescope mount, or panoramic telescope exceeds the tolerance authorized on any of the tests outlined, the piece and/or panoramic telescope must be sent to ordnance for adjustment.

b. Basic periodic tests are performed by the section under the supervision of the battery executive and the artillery mechanic. These tests are performed at the discretion of the unit commander. Suggested times for performance are once each year if the piece is used for nonfiring training; once every 3 months if the piece is fired; as soon as possible after intensive use, accidents, or traversing extremely rough terrain; and whenever the piece fires inaccurately for no readily apparent reason. The tests will reveal whether or not the on-carriage sighting equipment, the gunner's quadrant, and the fuze setter are in correct adjustment.

145. Preliminary Conditions

For the on-carriage equipment to be in correct adjustment, the following conditions must exist:

a. The line of sight of the panoramic telescope remains in a plane parallel to the vertical plane passing through the axis of the bore as the tube is elevated throughout its limits of elevation.

b. All indexes and scales read zero.

c. If the leveling-segment level bubbles and the leveling-mechanism level bubbles are centered, the telescope mount automatically compensates for error in azimuth caused by elevating the tube.

d. The sighting equipment is accurately bore sighted as described in paragraphs 128 through 143.

e. Prior to all tests of on-carriage fire control equipment, it is essential that the trunnions be leveled accurately. Leveling the trunnions is most easily accomplished as prescribed in paragraph 131. The best check to see that the trunnions are level is that the axis of the bore tracks a plumb line as described in paragraph 131*b*. If a plumb line cannot be used, approximate leveling may be accomplished with the gunner's quadrant atop the leveling plates on the breech ring.

f. Parallax shields for the panoramic telescope T149E1 and the telescope T159 must be prepared. In order to eliminate parallax in viewing a plumb line or a testing target, see paragraph 141.

Section II. TEST OF GUNNER'S QUADRANT

146. General

The gunner's quadrant must be in proper adjustment before conducting tests and adjustments of other sighting and fire control equipment. Inspect the shoes of the gunner's quadrant for dirt, nicks, or burs. Similarly, inspect the leveling plates on the upper surface of the breech ring and the quadrant seats on the quadrant mount. Dirt, nicks, or burs on these surfaces will cause the instrument to give inaccurate readings.

147. End-for-End Test

a. Set both the index arm and the micrometer scale of the gunner's quadrant at zero, making sure the auxiliary indexes match.

b. Place the quadrant on the leveling plates of the breech ring, the line of fire arrow pointing toward the muzzle, and center the quadrant bubble by turning the elevating handwheel.

c. Reverse the quadrant on the leveling plates (turn it end-for-end). If the bubble recenters, the quadrant is in adjustment and the test is completed.

d. If the bubble does not recenter, try to center it by turning the micrometer knob. If the bubble centers, read the black figures and divide by 2. This result is the correction. Place correction on the micrometer and level the tube by using elevation handwheel. Check again by reversing the quadrant. The bubble should center.

e. If the bubble does not center as in *d* above, move the radial arm down 1 graduation (10 mils) and perform the following operations: Turn the micrometer until the bubble centers; take the reading on the micrometer, add 10 to it, and divide the sum by 2; place this reading on the micrometer, leaving the arm at minus 10; level the bubble with elevation handwheel; and check by reversing quadrant on seats. The bubble should recenter. If the correction of error is more than plus or minus 0.4 mil, the quadrant must be adjusted by ordnance maintenance personnel.

148. Micrometer Test

a. Set the radial arm to read 10 mils on the elevation scale, and set the micrometer at 0.

b. Place the quadrant on the leveling plates on the breech ring, the line of fire arrow pointing toward the muzzle, and center the quadrant bubble by elevating the tube.

c. Set the radial arm at 0 on the elevation scale, and set the micrometer at 10 mils.

d. Reseat the quadrant on the leveling plates. The bubble should center.

Caution: Do not disturb the laying of the tube.

e. If the bubble does not center, the micrometer is in error and must be adjusted by ordnance maintenance personnel.

149. Comparison Test

Compare readings taken at low, medium, and high elevations with all of the gunner's quadrants of a battery on the quadrant seats of a *single* piece. The trunnions of this piece should be level. Any quadrant differing from the average by more than 0.4 mil at any elevation should be sent to an ordnance maintenance unit for adjustment.

150. Correction

When a gunner's quadrant requires a correction as determined by the end-for-end test, this correction is not carried during firing, but it is recorded and applied only when tests are being made.

Section III. TESTS FOR TELESCOPE MOUNT T179E2 AND PANORAMIC TELESCOPE T149E1

151. Purpose

The purpose of the tests for the telescope mount T179E2 and the panoramic telescope T149E1 is to determine whether the automatic azimuth compensating mechanism of the telescope mount actually establishes the tube (regardless of cant) in the correct vertical plane at all elevations. These tests check the adjustment and mounting of the panoramic telescope mount, the setting of the leveling-segment level and the leveling-mechanism level vials, and the alinement of the telescope socket. The test of the telescope mount described in paragraph 152 may be performed with the trunnions either level or canted. It reflects total errors of the entire mechanism. Because compensating errors of various parts of the mount may result in the weapon's testing out properly with this test, the other tests specified in paragraphs 153 through 156 must be performed regardless of the result of the test in paragraph 152. Total errors found in this test may then be reduced to errors in specific components.

152. Test of Telescope Mount T179E2

a. With bore sights in place and tube at a low elevation, traverse the tube so that the line of sight through the tube is on the plumb line; level the telescope mount by centering both the leveling-segment level and leveling-mechanism level bubbles.

b. Place the intersection of the cross hairs of the panoramic telescope reticle on any sharply defined aiming point and note the deflection.

c. Elevate the tube from minimum to maximum elevation (or limit of the plumb line) in 100-mil steps. At each step, traverse the tube (if necessary) to bring the line of sight back on the plumb line. Re-level the telescope mount in both directions and check for deviation of the line of sight from the aiming point. If the vertical cross hair is off the aiming point, it is turned to the aiming point with the azimuth micrometer. If the horizontal cross hair is off the aiming point, it is brought to the aiming point with the leveling knobs, and the bubble displacement is noted.

d. If the vertical cross hair deviates from the aiming point by more than ½ mil from the original deflection at any elevation tested or the correction for the deviation of the horizontal cross hair causes either of the bubbles to travel in excess of ½-vial graduation, the telescope mount is out of adjustment or improperly mounted. The weapon must be referred to ordnance maintenance personnel for adjustment or correction.

153. Test of Cross-Level Setting, Telescope Mount T179E2

a. Level the telescope mount T179E2 in both directions by centering the level bubbles.

b. Set the line of sight of the panoramic telescope at 0 with the parallax shield in place.

c. Suspend a plumb line to coincide with the vertical cross hair of the telescope reticle.

d. Turn the elevation knob of the panoramic telescope through the entire range of movement. If the line of sight deviates from the plumb line by more than ½ mil, the level vials are out of adjustment and must be adjusted by ordnance maintenance personnel.

154. Test of Longitudinal-Level Setting, Telescope Mount T179E2

a. Level the telescope mount T179E2 in both directions by centering the level bubbles.

b. By turning the azimuth micrometer knob of the panoramic telescope, set the line of sight to 1,600 mils with the parallax shield in place.

c. Suspend a plumb line to coincide with the vertical cross hair of the panoramic telescope.

d. Turn the elevation knob of the panoramic telescope through the entire range of movement. If the line of sight deviates from the plumb line by more than 1 mil, adjustment of the level vials is necessary. This adjustment must be performed only by ordnance maintenance personnel.

155. Test of Elevation Counter Synchronization

Using a gunner's quadrant that has been checked

for accuracy, measure the elevation of the tube at 0, 225, 625, 1,025, and 1,155 mils by placing the gunner's quadrant on the leveling plates and leveling the bubble of the gunner's quadrant. Turn the elevation counter setting knob until the vernier indexes are alined at each elevation. Check the readings of the elevation counter against the readings of the gunner's quadrant. If the 2 readings do not agree within ½ mil at *0* elevation and *1* mil at each of the other readings, the elevation linkage is out of adjustment and must be referred to ordnance maintenance personnel for adjustment.

156. Test for Panoramic Telescope T149E1

a. Zero the scales on the panoramic telescope T149E1.

b. Traverse and elevate the tube as necessary to place the panoramic telescope reticle cross hairs on an aiming point.

c. Rotate the telescope head through a complete circle (6,400 mils). The telescope cross hairs should return to the aiming point within 1 mil.

Section IV. FUZE SETTERS

157. General

Examine the stop which fits into the slot in the movable time ring and the adjusting pawl which engages the notch in the fixed fuze ring to see that their edges are not burred or bent. Depress the adjustable pawl against its spring to see that the movement of the pawl is free. In the following tests, be sure to test the fuze setter with the fuze for which it is designed; the time scale on the fuze setter must have the same graduations as the time ring on the fuze.

158. Time Scale Test

Set any convenient time on the scale. Test the time scale of the fuze setter by setting several fuzes.

159. Precaution

a. Before setting a fuze, make sure the "T" of the fuze setter is tight, to prevent any slipping of the scale indexes when the handle of the fuze setter is rotated. The time set on the fuze should agree with the time setting on the fuze setter within one-fourth of the smallest graduation on the fuze time ring. The tolerance amounts to 0.05 second for fuzes having 0.2-second graduations and 0.125 second for fuzes having 0.5-second graduations. If the fuzes set do not agree with the time set on the fuze setter, repeat the test as a check with a different setting. If the fuzes set still do not agree with the fuze setter, refer the instrument to an ordnance maintenance unit for adjustment.

b. Do not set any one live fuze more than twice. The fuze from a dud must never be used. Reset all fuzes to SAFE and replace the safety wire or cotter pin.

CHAPTER 10

MAINTENANCE AND INSPECTIONS

160. General

Maintenance and inspection are essential to insure that the section is prepared to carry out its mission immediately. Systematic maintenance and inspection drills provide the best insurance against unexpected breakdown at the critical moment when maximum performance is essential.

161. Disassembly, Adjustment, and Assembly

Disassemblies and adjustments of the weapon authorized to be performed by battery personnel are prescribed in TM 9-7212 and TM 9-7220, supplemented by instructions contained in Department of the Army supply manuals. No deviation from these procedures is permitted unless authorized by the responsible ordnance officer.

162. Records

a. The principal records pertaining to the weapon are the weapon record book (DA Form 9-13 and DA Form 9-13-1), a field report of accidents (AR 385-63), and the unsatisfactory equipment report (DA Form 468). Information on the purpose and use of these records may be found in the records themselves.

b. The principal records pertaining to the motor carriage and the section vehicle are the vehicle technical manual, lubrication order, accident report (Standard Form 91), and trip ticket (DD Form 110). Information on the purpose and use of these records may be found in the records themselves.

c. The chiefs of sections and the battery executive also should keep semipermanent records on the weapons and vehicles for information and guidance.

163. Maintenance

For detailed instructions concerning maintenance of the 155-mm gun M53, self-propelled, see TM 9-7212. For instructions concerning maintenance of the 8-inch howitzer M55, self-propelled, see TM 9-7220.

164. Inspections

Regular inspections are required to insure that materiel is maintained in serviceable condition.

a. The chief of section is responsible for the equipment within his section. He should inspect it thoroughly each day. If he sees the need for repair or adjustment, he notifies the executive immediately so that the necessary action may be taken.

b. The battery executive, accompanied by the artillery mechanic, should make a daily spot check inspection. He inspects different parts of the weapons and carriages each day to insure complete coverage every few days. At least once a month, the battery executive makes a thorough mechanical inspection of weapons, motor carriages, auxiliary equipment, tools, and spare parts.

c. Battery, battalion, and higher commanders should make frequent command inspections to assure themselves that the equipment in their commands is being maintained at prescribed standards of appearance, condition, and completeness.

d. For details on inspecting the 155-mm gun M53, see TM 9-2810 and TM 9-7212. For details on inspecting the 8-inch howitzer M55, see TM 9-2810 and TM 9-7220.

e. Duties of individuals in performing the necessary inspections and maintenance of the utility vehicle, weapon, and carriage are given in table VI. Work will be made routine, thorough, and rapid by following the drills outlined in table VI. When the section is reduced in strength, the chief of section must reassign duties to insure that all maintenance steps are completed.

165. Duties in Inspection Before Operation (March)

The inspection performed before operation is a final check on materiel prior to leaving the motor park for training in the field or the bivouac area for combat or before displacement. Bore sighting is accomplished at this inspection, if time permits. After inspection, and when all deficiencies have been corrected, the weapon and carriage are ready to go into action. For duties of section personnel, see table VI.

166. Duties in Inspection During Operation (March)

The inspections performed during operation are constant checks on the functioning of the vehicles and the security of all stowed equipment. The responsibilities and duties of section personnel are as follows:

a. The chief of section and the ammunition specialist supervise march discipline of the motor carriage and section vehicle, respectively, and assist the drivers in detecting obstacles that would cause injury to personnel or damage to the vehicles.

b. The gunner and other numbered cannoneers inspect security of stowed equipment and act as air sentinels as directed by the chief of section.

c. The drivers operate their respective vehicles and inspect all instruments and controls.

167. Duties in Inspection During Halt

The inspection at the halt is made to insure that the motor carriage, weapon, and section vehicle are in satisfactory operational condition. The halt provides the section with an opportunity to inspect for malfunctions that could not be detected during operation. Each member of the section will check those items listed in table VI, as appropriate.

168. Duties in Inspection and Maintenance After Operation

Immediately after operation, the motor carriage, weapon, and section vehicle are given whatever servicing and maintenance is needed to prepare them for further sustained action or to determine the need for maintenance by higher echelons. Bore sighting is accomplished, if time permits. These operations may be performed in the motor park, bivouac area, or combat position. Individual duties of gun (howitzer) section personnel are listed in table VI.

169. Duties in Weekly Inspection and Maintenance

In garrison, inspection and maintenance duties are

performed weekly; on maneuver or in combat, they are performed after each field operation.

a. Chief of Section. The chief of section supervises the section in weekly inspection and maintenance of gun (howitzer), motor carriage, section vehicle, tools, accessories, and equipment (TM 9-7212, TM 9-7220, LO 9-7212, and LO 9-7220). He obtains assistance of the artillery mechanic and battery mechanic for operations requiring skill and tools beyond the capabilities of the section.

b. Gunner and Cannoneers. The gunner and cannoneers perform normal maintenance as directed by the chief of section.

c. Drivers. The drivers perform normal preventive maintenance service in accordance with TM 9-7212 and TM 9-7220.

CHAPTER 11

DECONTAMINATION OF EQUIPMENT

170. General

Equipment which has been contaminated by chemical, biological, or radiological agents constitutes a danger to personnel. *Contamination* means the spreading of an injurious agent in any form and by any means. Persons, objects, or terrain can be contaminated. *Decontamination* is the process of making any contaminated place or thing safe for unprotected personnel. This can be done by covering, removing, destroying, or changing into harmless substances the contaminating agent or agents. Generally, only equipment contaminated by persistent agents need be decontaminated.

171. Decontamination for Chemical Agents

a. Ammunition. With rags, wipe off visible contaminant from projectiles. Apply DANC (decontamination agent, noncorrosive, M4), wipe with solvent-soaked rag, and then dry. If DANC is not available, scrub with soap and cool water. Slurry (equal weights of water and chloride of lime) can be used on contaminated ammunition containers, but it must not be allowed to penetrate into the ammunition itself.

b. Instruments. If exposed to corrosive gases, clean instruments as soon as possible with solvent, allow them to aerate, and apply a thin coat of light machine oil. A rag dampened with DANC may be used, followed by drying with a clean rag and then applying a coat of machine oil. DANC injures clear plastic or hard rubber surfaces.

c. Weapons. Remove dirt, dust, grease, and oil from weapons. Do not apply wet mix but allow surfaces to air after oil and dirt have been removed. DANC can be used on all metal surfaces except the bore. Also effective on metal are hot water and soap or cleaning solvent. After decontamination, weapons are dried and oiled.

d. Automotive Equipment. Exposure to the air can neutralize light contamination from spray. For heavier contamination, use DANC on interior or exterior surfaces that personnel are likely to touch. For larger area decontamination, wash vehicle with water and scrub painted surfaces with soap and water.

172. Decontamination for Biological and Radiological Agents

a. General. After a contaminating attack, recovery of equipment may be achieved either by waiting, to permit the decay of contamination, or by active decontamination, to reduce the danger to a level where it is no longer a significant hazard to operating personnel. Decontamination may be either rough or detailed, depending on the urgency of the military situation. The procedure adopted will be a command decision.

b. Rough Decontamination. Rough decontamination is performed when urgency is the main factor. Its purpose is to reduce contamination sufficiently to permit personnel to work with, or close to, equipment for limited periods. Rough decontamination may be achieved by means of water or steam, if available. Soap or other detergent used in conjunction with water or steam aids in decontamination.

c. Detailed Decontamination. Detailed decontamination, in which the emphasis is on thoroughness, will be carried out in rear areas and repair bases and includes procedures of surface decontamination, aging, sealing, and disposal.

173. References

For further information on decontamination, see FM 21-40, TM 3-220, and TF 3-1407.

CHAPTER 12

DESTRUCTION OF EQUIPMENT

174. General

a. Tactical situations may arise in which it is necessary to abandon equipment in the combat zone. In such a situation, all abandoned equipment must be destroyed to prevent its use by the enemy.

b. The destruction of equipment subject to capture or abandonment in the combat zone will be undertaken only upon authority delegated by a division or higher commander.

175. Plans

All batteries will prepare plans for destroying their equipment in order to reduce the time required should destruction become necessary. The principles to be followed are—

a. Plans for destruction of equipment must be adequate, uniform, and easily carried out in the field.

b. Destruction must be as complete as the available time, equipment, and personnel will permit. Since complete destruction requires considerable time, *priorities* must be established so that the more essential parts are destroyed first.

c. The same essential parts must be destroyed on all like units to prevent the enemy from constructing a complete unit from undamaged parts.

d. Spare parts and accessories must be given the same priorities as the parts installed on the equipment.

176. Methods

To destroy equipment adequately and uniformly, all personnel of the unit must know the plan and priority of destruction and be trained in the methods of destruction.

177. References

For detailed information on destruction of the 155-mm gun T80, 8-inch howitzer T89, fire control equipment, and the motor carriage M53 or M55, see TM 9-7212 and TM 9-7220; for destruction of ammunition, see TM 9-1901; for destruction of the cargo trailer, see the technical manual for that particular vehicle.

CHAPTER 13

SAFETY PRECAUTIONS

178. General

Safety precautions to be observed in training are prescribed in AR 385-63. Additional information is found in FM 6-40, FM 6-140, TM 9-1901, TM 9-7212, and TM 9-7220. The more important safety precautions are summarized in paragraphs 179 through 182.

179. Ammunition

a. All ammunition on the ground at the firing position must be so placed that it is protected against explosion in case of accident at the piece position. Fire and explosive or flammable materials must be kept away from ammunition. Ammunition should be protected from direct rays of the sun by use of a tarpaulin or other suitable covering.

b. Battery personnel must not attempt to disassemble fuzes.

c. If for any reason a round is not fired after the time fuze has been set, the fuze must be reset to SAFE before it is restowed. M514-series VT fuzes must be reset to initial setting as shipped.

d. All rounds not fired which have been prepared for firing must be checked by the chief of section to insure that all powder increments are present in proper order and condition and that they are of the proper lot number. For further details, see FM 6-40 and FM 6-140.

180. Misfires

a. In the event of a misfire, two more attempts are made to fire the piece.

Caution: The piece should remain as laid and all personnel must stay clear of the muzzle and path of recoil. All personnel not required for the operation should be cleared from the vicinity.

b. If the primer is heard to fire, a minimum of 10 minutes will be allowed before the breech is opened and the faulty charge removed. The faulty charge, as well as the unused increments and igniter pads, must be stored separately from other charges and disposed of (see TM 9-1900 for further details).

c. If the primer is not heard to fire, two more attempts to fire will be made. If, after the third attempt, the primer is not heard to fire, proceed as follows:

    (1) Wait 2 minutes. If the primer can be removed by No. 2 standing clear of the path of recoil, the primer may be removed and a new one inserted.

    (2) If the primer cannot be removed safely as described in (1) above, no attempt will be made to open the breech or replace the primer for 10 minutes.

d. Misfire primers should be handled carefully and disposed of quickly since there is a possibility of a primer hangfire.

181. Drill and Firing

a. The piece is kept unloaded except when firing is imminent.

b. Members of the section on the ground will pass in rear of the carriage when they go from side to side.

c. Personnel must stay a safe distance from the breech to prevent injury when the piece recoils.

d. During firing, personnel should use cotton in their ears to protect eardrums against injury.

182. Safety Officer

In training, there must always be a safety officer for each artillery unit firing. For duties of safety officer, see FM 6–40.

CHAPTER 14

TRAINING

183. Purpose and Scope

The purpose of this chapter is to present the requirements for training section personnel in the performance of their duties in service of the piece. It includes general information on the conduct of training and tests to be given for the qualification of gunners.

184. Objectives

The objectives of training are to speed the attainment of proficiency by cannoneers in their individual duties and, through drill, to weld them into an effective, coordinated team that is able to function efficiently in combat. During training, the supervisor should keep in mind the proficiency sought by the appropriate Army Training Tests (ATT). Maximum efficiency is attained through regular drills.

185. Conduct of Training

a. Training will be conducted in accordance with the principles set forth in FM 21-5. The goal of training should be the standards set forth in FM 6-125, TM 6-605, and AR 611-201.

b. In general, individual training is conducted by noncommissioned officers as far as practicable. Officers are responsible for preparing training plans, for conducting unit training, and for supervising and testing individual training.

c. Throughout training, the application of prior instructions to current training must be emphasized.

d. A record of the training received by each individual should be kept on a progress card which may be maintained by each chief of section for each man in his section. This card should show each period of instruction attended, tests taken, and remarks pertaining to progress. Progress cards should be inspected frequently by the battery executive to make sure they are being kept properly and to determine the state of training. *Requiring the chief of section to keep these records emphasizes his responsibility toward his section.*

e. The necessity for developing leadership and initiative in noncommissioned officers must be emphasized constantly throughout training.

186. Standard to be Attained

A satisfactorily trained weapon section must be capable of performing the following functions in the times indicated (TM 6-605):

a. Firing. The section must fire 10 rounds (drill ammunition) at different deflections, elevations, and time fuze settings, using the same charge, in 5.5 minutes by day and 6 minutes by night for the 155-mm gun section and in 10 minutes by day and 15 minutes by night for the 8-inch howitzer section. Changes in data should be typical for a time bracket adjustment; data are announced from prepared cards.

b. After-Firing Care and Maintenance of Armament. The piece being in position, the section must clean and lubricate, disassemble and assemble the breech and firing mechanism, and inspect the weapon in 45 minutes by day and 60 minutes by night. All tools and materials required should be available at the position.

c. Six-Month Inspection and Maintenance of Armament. The weapon prepared for action in the gun park, the section must clean and lubricate all parts and assemblies, as authorized, and prepare for ordnance inspection in 3 hours 10 minutes. All tools and materials required should be available in the gun park.

d. Section Drill. Each member of the section should know the duties of all other members of the section and be able to perform efficiently in all positions. See paragraphs 187 through 244 for tests to be given for the qualification of gunners.

CHAPTER 15

TESTS FOR QUALIFICATIONS OF GUNNERS

Section I. GENERAL

187. Purpose and Scope

This section prescribes the tests to be given in the qualification of gunners. The purposes of the tests are—

a. To provide a means of determining the relative proficiency of the individual artillery soldier in the performance of the duties of the gunner. *The tests will not be a basis for determining the relative proficiency of batteries or higher units.*

b. To serve as an adjunct to training.

188. Standards of Precision

The candidate will be required to perform the tests in accordance with the standards listed in *a* through *d* below.

a. Scale settings must be exact and matching indexes must be brought into coincidence.

b. Level bubbles must be exactly centered.

c. The vertical cross hair in the reticle of the panoramic telescope must be alined on the left edge of the aiming post or on exactly the same part of the aiming point each time the piece is laid.

d. Final motions of azimuth and elevation setting knobs, as well as traversing and elevation handwheels, must be made in the appropriate direction (par. 118).

189. Assistance

The candidate will receive no unauthorized assistance. Each candidate may select authorized assistants as indicated in the tests. In the event a candidate fails any test because of the fault of the examiner or any assistant, the test will be disregarded, and the candidate will be given another test of the same nature.

190. Time

The time for any test will be the time from the last word of the command to the last word of the candidate's report. The candidate may begin any test after the first word of the first command and should not be charged for any time used by the examiner.

191. Scoring

Scoring will be conducted in accordance with the two paragraphs PENALTIES and CREDITS under each subject. If a test is performed correctly, credit will be given in accordance with the paragraph CREDIT under each subject. No credit will be allowed if conditions exist as specified in the paragraph PENALTIES. No penalty will be assessed in excess of the maximum credit for each test.

192. Preparation for Tests

The piece will be prepared for action and the candidate posted at the proper position corresponding to the test being conducted or as indicated in the SPECIAL INSTRUCTIONS paragraphs under each subject. The examiner will insure that the candidate understands the requirements of each test and will require the candidate to report "I am ready," before each test.

193. Qualification Scores

Minimum scores required for qualification in the courses are as follows:

Individual classification	Points
Expert gunner	90
First-class gunner	80
Second-class gunner	70

194. Outline of Tests

Section No.	Subject	No. of tests	Points each	Maximum credit
II	Direct laying, direct fire telescope.	4	2	8
III	Indirect laying, deflection only.	18	2	36
IV	Displacement correction.	1		4
	Part I		(3)	(3)
	Part II		(1)	(1)
V	Measuring deflection.	2	4	8
VI	Laying for elevation, elevation counter.	3	2	6
VII	Laying for elevation, gunner's quadrant.	3	2	6
VIII	Measuring elevation.	1	5	5
IX	Measuring angle of site to mask.	1	4	4
X	Sighting and fire control equipment.	2	4	8
XI	Materiel.	3	5	15
	Total credit			100

Section II. TEST, DIRECT LAYING, DIRECT FIRE TELESCOPE

195. Scope of Tests

Four tests (2 groups of 2 tests each) will be conducted in which the candidate will be required to execute commands similar to those given in paragraph 197. Tests 1 and 2 (and tests 3 and 4) will be executed as one series of commands.

196. Special Instructions

a. A stationary target will be placed approximately 600 yards from the piece.

b. The coarse azimuth and micrometer scale will be set at zero and indexes will be matched.

c. The candidate will be posted as the gunner.

d. The piece will be pointed so that a shift of approximately 100 mils will be required for tests 1 and 3, and it will not be necessary to shift the motor carriage for any of the 4 tests.

e. Laying at the termination of tests 1 and 3 will not be disturbed at the beginning of tests 2 and 4.

f. The examiner will announce the assumed direction of the movement of the target before tests 1 and 3. The assumed direction of the movement of the target in test 3 will be opposite to that in test 1.

197. Outline of Tests

Test No.	Examiner commands (for example)	Action of candidate
1 and 3	TARGET, THAT TANK, FROM LEFT TO RIGHT, LEAD 5, RANGE 600.	Centers cant corrector level bubble. Traverses tube until proper lead has been established. Places proper range line of reticle on the center of the visible mass of the target. Checks cant corrector level. Gives command FIRE, when ready, and steps clear.
2 and 4	RIGHT (LEFT) 6, ADD (DROP) 200.	Same as test 1 above.

198. Penalties

No credit will be allowed if, after each test—

a. The azimuth scale has been moved from zero.

b. The indexes on the azimuth micrometer have been moved from zero.

c. The cant corrector bubble is not centered.

d. The lead in mils is not set properly.

e. The proper range line of the reticle is not on the center of the visible mass of the target.

199. Credit

Time in seconds, exactly or less than 8 $8\frac{2}{5}$ 9
Credit -- 2.0   1.5   1.0

Section III. TEST, INDIRECT LAYING, DEFLECTION ONLY

200. Scope of Tests

Eighteen tests will be conducted in which the candidate will be required to execute commands similar to those given in paragraph 202. Tests 1 through 4 (and tests 5–9, 10–13, and 14–18) will be executed as one series of commands.

201. Special Instructions

a. Commands will not necessitate movement of motor carriage.

b. The examiner will select a suitable aiming point and identify it to the candidate.

c. Commands for special corrections will be given *only* in the tests indicated in the examples given in paragraph 202.

d. The command for new deflections for each test will be within the following prescribed limits:

Test No.	Maximum change (mils)	Minimum change (mils)
2 and 11	180	140
3 and 12	90	70
4 and 13	40	20
7 and 16	100	60
8 and 17	50	30
9 and 18	20	10

e. The piece will be laid with the correct setting at the conclusion of each test before proceeding with the next test.

f. For these tests, aiming posts will be set out at prescribed deflection and distances.

g. The examiner will designate the section number of the piece to be used and will announce, when applicable, special corrections in deflection to be applied by the candidate.

h. The candidate will be posted as the gunner.

202. Outline of Tests

Test No.	Examiner commands (for example)	Action of candidate
1 and 10	SPECIAL CORRECTIONS, DEFLECTION 3290. NO. 1, LEFT 7.	Sets deflection and applies special correction. Centers leveling bubbles. Traverses piece until vertical cross hair is on left edge of aiming posts. Checks centering of level bubbles. Re-lays if necessary. Calls "Ready" and steps clear.
2 and 11	DEFLECTION 3153.	Sets deflection. Lays on aiming posts. Checks centering of level bubbles. Re-lays if necessary. Calls "Ready" and steps clear.
3 and 12	DEFLECTION 3236. NO. 1, RIGHT 4. At conclusion of test 4 (13) give CEASE FIRING, END OF MISSION. (No time considered for this operation.)	Same as test 2 above.
4 and 13		Same as test 2 above.

Test No.	Examiner commands (for example)	Action of candidate
5 and 14	AIMING POINT, CHURCH STEEPLE (or such-and-such), REFER.	Refers telescope to church steeple. Reads deflection on coarse azimuth and micrometer scales and calls "No. 1, deflection (so much)."
6 and 15	DEFLECTION 3000, REFER.	Sets deflection on azimuth counter. Verifies that vertical cross hair of the reticle is on appropriate part of church steeple. Calls "No. 1, deflection 3000." Steps clear.
7 and 16	SPECIAL CORRECTIONS, DEFLECTION 3080. NO. 1, LEFT 7.	Same as test 1 above.
8 and 17	DEFLECTION 3120.	Same as test 2 above.
9 and 18	DEFLECTION 3135.	Same as test 2 above.

203. Penalties

No credit will be allowed if, after each test—

a. The deflection is not set correctly.

b. The level bubbles are not centered.

c. The vertical cross hair of the telescope is not on the aiming point or left edge of aiming posts, as the case may be.

d. The last motion of the traverse was not made to the right.

204. Credit

Time in seconds, exactly or less than—

Tests 1, 10, 7, and 16, each	12	13	14
Other tests, each	8	9	10
Credit	2.0	1.5	1.0

Section IV. TEST, DISPLACEMENT CORRECTION

205. Scope of Test

One test, consisting of two parts, will be conducted in which the candidate will be required to execute the commands given in paragraph 207.

206. Special Instructions

a. Aiming posts will be set out at prescribed distances.

b. An assistant, selected by the candidate, will be stationed near the far aiming post.

c. The examiner will require the candidate to lay the piece on an announced deflection and report, "I am ready."

d. The motor carriage or the far aiming post will then be moved so that an aiming post displacement of 5 to 10 mils occurs.

e. The laying of the piece at the termination of part I will not be disturbed for part II.

207. Outline of Test

a. Part I.

Examiner commands	Action of candidate
CORRECT FOR DISPLACEMENT	Lays the piece so that the far aiming post appears midway between the near aiming post and the vertical cross hair of the telescope. Checks centering of level bubbles. Re-lays if necessary. Calls "Ready" and steps clear.

b. Part II.

Examiner commands	Action of candidate
ALINE AIMING POSTS.	Records deflection on breech and announces "Deflection (so much) recorded." Directs assistant in alining aiming posts. Calls "Ready" and steps clear.

208. Penalties

No credit will be allowed for either part if—

a. Part I.
  (1) The far aiming post does not appear midway between the near aiming post and the vertical cross hair of the telescope.
  (2) The bubbles are not centered.
  (3) The last motion of traverse was not made to the right.

b. Part II.
  (1) The deflection is other than the announced deflection.

(2) The aiming posts are not properly alined.

(3) The vertical cross hair of the telescope reticle is not on the left edge of the aiming posts.

209. Credit

Part I, time in seconds, exactly or less than	3	3⅓	3⅔	4
Credit	3.0	2.0	1.5	1.0
Part II, no time limit				
Credit	1.0	----	----	----

Section V. TEST, MEASURING DEFLECTION

210. Scope of Test

Two tests will be conducted in which the candidate will be required to measure and report a deflection in accordance with the commands given in paragraph 212.

211. Special Instructions

a. The piece will be laid on aiming posts to the right front.

b. The examiner will select 2 aiming points: the aiming point for test 1 will be within 200 mils to the left or right of the aiming posts, and the aiming point for test 2 will be within 200 mils on the opposite side of the aiming posts.

c. The appropriate aiming point will be designated by the examiner and identified by the candidate prior to the start of each test.

212. Outline of Tests

Test No.	Examiner commands	Action of candidate
1	NO. 1, AIMING POINT, CHURCH STEEPLE TO LEFT FRONT, REFER.	Centers the level bubbles. Refers to aiming point. Checks centering of bubbles. Reads deflection on coarse azimuth and micrometer scales and reports, "No. (so-and-so) deflection (so much)" and steps clear.
2	NO. 1, AIMING POINT, WATER TOWER, RIGHT FRONT, REFER.	Same as test 1 above.

213. Penalties

No credit will be allowed if—

a. The level bubbles are not centered properly.

b. The vertical cross hair of the telescope reticle is not on the aiming point properly.

c. The deflection is not announced correctly.

d. The traversing handwheel is turned.

214. Credit

Time in seconds, each test,
exactly or less than.............. 5   5⅗   6   6⅗
Credit 4.0   3.0   2.0   1.5

Section VI. TEST, LAYING FOR ELEVATION, ELEVATION COUNTER

215. Scope of Test

Three tests will be conducted in which the candidate will be required to execute commands similar to those given in paragraph 217.

216. Special Instructions

a. Each test will require a change of settings and the accompanying laying of the piece in elevation. (All commands given will be within the limits of 200 to 400 mils on the elevation counter.)

b. Commands for elevation for tests 2 and 3 will not be made in multiples of 5 mils.

217. Outline of Tests

Test No.	Examiner commands	Action of candidate
1	ELEVATION 290.	Sets announced elevation. Matches elevation vernier. Checks level bubbles. Calls "Ready."
2	ELEVATION 326.	Same as test 1 above.
3	ELEVATION 323.	Same as test 1 above.

218. Penalties

No credit will be allowed if after each test—

a. The elevation counter is not set accurately.

b. The elevation vernier is not matched.

c. The last motion of the tube was not in the direction in which it is most difficult to turn the elevating handwheel.

219. Credit

Time in seconds, exactly or less than	$6\tfrac{3}{5}$	$7\tfrac{3}{5}$	$8\tfrac{3}{5}$
Credit	2.0	1.5	1.0

Section VII. TEST, LAYING FOR ELEVATION, GUNNER'S QUADRANT

220. Scope of Tests

Three tests will be conducted in which the candidate will be required to execute commands similar to those given in paragraph 222.

221. Special Instructions

a. The gunner's quadrant will be set at zero for the first test.

b. Each succeeding test will require a change of quadrant setting within the limits of 30 to 60 mils.

c. The candidate will be posted to the left of and facing the breech with the gunner's quadrant in his hand.

d. An assistant, selected by the candidate, will be posted to operate the elevating handwheel.

222. Outline of Tests

Test No.	Examiner commands	Action of candidate
1	QUADRANT 190.	Sets elevation on gunner's quadrant and seats quadrant. Has assistant elevate or depress the tube until quadrant bubble is centered. Calls "Ready," and waits for examiner to verify laying.
2	QUADRANT 245.	Same as test 1 above.
3	QUADRANT 215.	Same as test 1 above.

223. Penalties

No credit will be allowed if, after each test—

a. The quadrant elevation is not set correctly.

b. The quadrant is not properly seated.

c. The quadrant bubble is not properly centered.

224. Credit

Time in seconds, exactly or less than 6   6⅗   7
Credit --- 2.0   1.5   1.0

Section VIII. TEST, MEASURING ELEVATION

225. Scope of Test

One test will be conducted in which the candidate will be required to measure the elevation by means of the gunner's quadrant.

226. Special Instructions

Prior to the test the examiner will lay the tube at a selected elevation, measure the elevation, and then set the gunner's quadrant at zero.

227. Outline of Test

Examiner commands	Action of candidate
MEASURE THE ELEVATION.	Places gunner's quadrant on the quadrant seats of the breech ring. Levels bubble by raising or lowering the index arm and turning the micrometer knob. Announces "No. (so-and-so) elevation (so much)," and hands quadrant to the examiner.

228. Penalties

No credit will be allowed if—

a. The quadrant bubble is not centered when the quadrant is seated properly.

b. The elevation is not announced correctly.

229. Credit

Time in seconds, exactly or less than	8	$9\tfrac{2}{5}$	$10\tfrac{3}{5}$
Credit	5.0	3.5	2.0

Section IX. TEST, MEASURING ANGLE OF SITE TO MASK

230. Scope of Test

One test will be conducted in which the candidate will be required to execute the command given in paragraph 232.

231. Special Instructions

a. The piece, prepared for action, will be placed 200 to 400 yards from a mask of reasonable height.

b. The tube will be pointed so that it is 100 to 150 mils above the crest and 100 to 150 mils right or left of the highest point of the crest.

c. The candidate will be posted at the left rear of the breech with the gunner's quadrant in his hand.

d. An assistant, selected by the candidate, will be posted as gunner to elevate or depress and traverse the tube as directed by the candidate.

232. Outline of Test

Examiner commands	Action of candidate
MEASURE ANGLE OF SITE TO MASK.	Sights along lowest element of the bore, and has the tube moved until the line of sight just clears the highest point of the crest. Rotates the elevation counter setting knob until the vernier is alined and elevation counter is locked. Checks centering of level bubbles. Reads elevation from elevation counter. Reports, "No. (so-and-so), angle of site to the mask (so much)."

233. Penalties

No credit will be allowed if—

a. The line of sight along the lowest element of the bore does not just clear crest.

b. The quadrant bubble is not centered when the quadrant is properly seated.

c. The angle of site is not announced correctly.

234. Credit

Time in seconds, exactly or less than	15	16	17	18
Credit	4.0	3.0	2.0	1.5

Section X. TEST, SIGHTING AND FIRE CONTROL EQUIPMENT

235. Scope of Tests

Two tests will be conducted in which the candidate will be required to demonstrate the methods employed in making the prescribed tests and authorized adjustments or to describe the action taken (i.e., send to the ordnance maintenance company) if adjustment is not authorized to be made by using personnel.

236. Special Instructions

The piece will be prepared for action with trunnions level and the tube in center of traverse.

237. Outline of Tests

Test No.	Examiner commands	Action of candidate
1	PERFORM END-FOR-END TEST ON GUNNER'S QUADRANT.	Performs test as prescribed in paragraph 147. Calls "Error (so many) mils, quadrant serviceable (unserviceable)" and hands quadrant to examiner for verification.
2	PERFORM MICROMETER TEST ON GUNNER'S QUADRANT.	Performs test as prescribed in paragraph 148. Calls "Quadrant micrometer is (is not) in error." States what action, if any, should be taken.

238. Penalties

a. General. The tests are not essentially speed tests. The purpose of the prescribed time limits is to insure that the candidate can perform the operation without wasted effort.

b. Test 1. No credit will be allowed if—

(1) The bubble of the gunner's quadrant does not center when verified by the examiner.

(2) The error (one-half of the amount of the angle which was indicated when the quadrant first was reversed and the bubble centered by moving the index arm and micrometer) is not announced correctly by the candidate.

(3) The candidate does not declare the quadrant unserviceable if the error (necessary correction) exceeds 0.4 mil or does not declare the quadrant serviceable if the error (necessary correction) is 0.4 mil or less.

(4) The time to complete the test exceeds 2 minutes.

c. Test 2. No credit will be allowed if—

(1) The procedure is not followed correctly.

(2) The time to complete the test exceeds 1 minute.

(3) The candidate fails to report necessary action to be taken.

239. Credit

If the tests and adjustments are performed correctly within the prescribed time limit, maximum credit will be given as follows:

Test 1	4
Test 2	4
Total	8

Section XI. TEST, MATERIEL

240. Scope of Tests

The candidate will be required to perform three tests as prescribed in paragraph 242.

241. Special Instructions

a. Tests 1 and 2. For tests 1 and 2, a paulin will be placed on the firing platform for the convenience of the candidate in laying out the disassembled parts. The candidate will be allowed to select the tools and accessories necessary for the performance of the tests prior to the start of the tests. The candidate may have an assistant to aid him in lowering and lifting the breechblock.

b. Test 3. A complete set of lubrication equipment authorized for use of battery personnel, including lubrication order, will be made available on a nearby paulin. Every type of lubricant used on the piece will be available in plainly labeled containers and placed on the paulin.

242. Outline of Tests

Test No.	Examiner commands	Action of candidate
1	DISASSEMBLE BREECH AND FIRING MECHANISM.	Performs the operation as described in TM 9-7212 or TM 9-7220, laying the disassembled parts on the paulin. After disassembly, identifies all parts to examiner.
2	ASSEMBLE BREECH AND FIRING MECHANISM.	Performs the operation as described in TM 9-7212 or TM 9-7220.

Test No.	Examiner commands	Action of candidate
3	PERFORM DAILY, WEEKLY, AND MONTHLY LUBRICATION TEST.	Using the lubrication order as a guide, selects proper lubrication equipment and lubricant and shows *how, when,* and with *which lubricant* each lubrication point on the piece is serviced (actual lubrication is not performed).

243. Penalties

a. The tests are not essentially speed tests. The purpose of the maximum time limits is to insure that the candidate can perform the operations without wasted effort.

b. No credit will be given if the following time limits are exceeded:

	155-mm gun	*8-inch howitzer*
Test 1	8 minutes	12 minutes
Test 2	12 minutes	16 minutes
Test 3	5 minutes	5 minutes

c. A penalty of one-half point will be assessed for each component part that is not correctly identified or omitted in test 1. There is no time limit imposed on the identification of component parts. However, the examiner may reduce the grade if it becomes obvious that the candidate is not familiar with the correct nomenclature.

d. A penalty of one-half point will be assessed for each lubrication point missed or lubricated improperly and for each time the proper lubricating device or proper lubricant is not selected.

244. Credit

a. The candidate will be scored on the general merit of his work in addition to the specific requirements above.

b. If each test is performed correctly within the prescribed time limit, maximum credit will be given as follows:

Test 1	5 points
Test 2	5 points
Test 3	5 points
Total	15 points

APPENDIX

REFERENCES

1. Miscellaneous Publications

AR 320–50	Authorized Abbreviations.
AR 385–63	Regulations for Firing Ammunition for Training, Target Practice, and Combat.
AR 600–70	Badges.
AR 611–201	Manual of Enlisted Military Occupational Specialties.
AR 700–38	Unsatisfactory Equipment Report (Reports Control Symbol CSGLD-247 (R2)).
AR 750–5	Maintenance Responsibilities and Shop Operations.
ATP 6–300	Field Artillery Unit Training Program.
ATT 6–1	Training Tests for Field Artillery Howitzer or Gun Battery.
ATT 6–2	Training Tests for Field Artillery Battalion Firing.
ATT 6–3	Training Tests for Field Artillery Rocket Battery and Battalion Tests.
DA Pamphlet 108–1	Index of Army Motion Pictures, Television Recordings, and Film Strips.
DA Pamphlet 310-series	Index of Military Publications.
SR 320–5–1	Dictionary of United States Army Terms.
TOE 6–437R	Field Artillery Battery, 155-mm Gun, Self-Propelled.
TOE 6–447R	Field Artillery Battery, 8-inch Howitzer, Self-Propelled.
TF 3–1407	Decontamination Procedures—Part I: Basic Techniques.

FM 5-15	Field Fortifications.
FM 5-20	Camouflage, Basic Principles.
FM 5-20B	Camouflage of Vehicles.
FM 5-20D	Camouflage of Field Artillery.
FM 5-25	Explosives and Demolitions.
FM 6-40	Field Artillery Gunnery.
FM 6-101	The Field Artillery Battalion.
FM 6-125	Qualification Tests for Specialists, Field Artillery.
FM 6-140	The Field Artillery Battery.
FM 17-50	Logistics, Armored Division.
FM 21-5	Military Training.
FM 21-30	Military Symbols.
FM 21-40	Defense Against CBR Attack.
FM 21-60	Visual Signals.
FM 22-5	Drills and Ceremonies.
FM 23-65	Browning Machine Gun, Caliber .50 HB, M2.
FM 25-10	Motor Transportation, Operations.
LO 9-7212	Gun, Self-Propelled, Full Tracked, 155-mm, T97.
TM 3-220	Decontamination.
TM 6-605	Field Artillery Individual and Unit Training Standards.
TM 9-575	Auxiliary Sighting and Fire Control Equipment.
TM 9-1590	Fuze Setters M14, M22, M23, M25, and M27.
TM 9-1900	Ammunition, General.
TM 9-1901	Artillery Ammunition.
TM 9-2300	Artillery Materiel and Associated Equipment.
TM 9-2810	Tactical Motor Vehicle Inspections and Preventive Maintenance Services.
TM 9-2853	Preparation of Ordnance Materiel for Deep Water Fording.
TM 9-6133	Ordnance Maintenance: Fuze Setter M26.
TM 9-7212	Self-Propelled 155-mm Gun T97.
TM 9-7220	Self-Propelled 8-Inch Full Tracked Howitzer M55 (T108).

TM 21-301	Driver Selection, Training and Supervision, Half-Track and Full-Track Vehicles.
TM 21-305	For The Wheeled Vehicle Driver.
TM 21-306	Manual for the Full-Track Vehicle Driver.

2. Supply Manuals

ORD 8 SNL G-259	Gun, Self-Propelled, Full-Tracked 155-mm, T97, and Howitzer, Self-Propelled, Full-Tracked 8-in. T108.
ORD 3 SNL K-1	Abrasives, Adhesives, Cleaners, Preservatives, Recoil Fluids, Special Oils, and Related Items.
SM 9-5-1315	Ammunition, 75-millimeter through 125-millimeter.
SM 9-5-1320	Ammunition, over 125-millimeter.
SM 9-5-1345	Land Mines and Components.
SM 9-5-1370	Pyrotechnics, Military, All Types.
SM 9-5-1375	Bulk Propellants and Explosive Devices.
SM 9-5-1390	Fuzes and Primers

3. Forms

DA Form 9-13	Weapon Record Book.
DA Form 468	Unsatisfactory Equipment Report.

4. Atomic Publications

FM 6-150(S)	Functions and Procedures for Atomic Field Artillery and Supporting Ordnance Units (U).
TC 6-6	Use of Shell, High Explosive, Spotting (HES), 8-inch, T347, in Conjunction with the Delivery of Shell, Atomic Explosive (AE), 8-inch T317.
TM 39-0-1(S)	Numerical Index to Joint Special Weapons Publications. (U)
TM 39-B33-1(S)	Classified.

INDEX

	Paragraphs	Page
Aiming:		
Points	119	70
Posts	120	71
Displacement correction	121	73
Ammunition:		
Care	126	78
Direct fire	100	57
Safety precautions	178	111
Basic periodic tests	144–159	92
Boresighting:		
Distant aiming point	138, 139	87
General	128–132	81
Standard angle method	140–143	87
Testing target method	133–137	83
Capabilities of motor carriage	3	5
Cease firing	123	76
Changes in data during firing	124	76
Composition of gun (howitzer) section	5	9
Decontamination:		
Biological and radiological	172	107
Chemical agents	171	106
General	170	106
References	173	108
Definition of terms	2	4
Destruction of equipment	174–177	109
Direct laying, firing by:		
Ammunition	100	57
Conduct of fire	99	56
Duties of:		
Chief of section	103–109	63
Gunner	110–116	66
Remainder of section	117	69
Field of fire	98	56

	Paragraphs	Page
Direct laying, firing by—Cont.:		
General	96	55
Preparation of range card	97	55
Trajectories	101	59
Vertical displacement table	102	61
Disassembly, adjustment, and assembly by battery personnel	161	101
Distant aiming point	119	70
Duties of personnel, general	6	9
Fuze setters	157–159	99
Gunner's quadrant:		
Basic periodic test	146–150	94
Laying for elevation	29	27
Indirect laying, firing by:		
Deflection	39	31
Description of individual duties:		
Ammunition handlers	88–90	53
Ammunition specialist	81–87	50
Chief of section	21–36	24
Drivers	91–95	53
Gunner	37–47	31
No. 1	48–52	57
No. 2	53–61	59
No. 3	62–66	41
No. 4	67–74	43
No. 5	75–80	47
General	19	22
General duties of individuals	20	22
Inspection:		
After operation	168	104
Before operation	165	103
During halt	167	104
During operation	166	103
Weekly	169	104
Laying for elevation:		
Elevation counter	41	34
Gunner's quadrant	29	27

	Paragraphs	Page
Lights, aiming posts	120	71
Maintenance, records	162	101
March order	17	21
Measuring:		
Angle of site to mask	23	25
Distance from piece to aiming posts	120	71
Elevation	30	28
Parallax	141	88
Plumb line	131	82
Precision in laying	118	70
Preparing the piece for:		
Firing	15, 16	20
Traveling	17, 18	21
References	4, app	8, 139
Safety precautions:		
Ammunition	179	111
Drill and firing	181	113
General	178	111
Misfires	180	112
Safety officer	182	113
Section data board	127	80
Section drill:		
Change posts	11	14
Dismount	13	17
Fall out	14	18
Form the section	9	12
General	7, 8	11
Mount	12	17
Post the section	10	14
Testing target, improvement	122	73
Test(s):		
Basic periodic	154–156	98
Fuze setters	157–159	99
Gunner's qualification:		
Assistance	189	118
Direct laying	195–199	120
Displacement correction	205–209	125

	Paragraphs	Page
Gunner's qualification—Cont.:		
Indirect laying, deflection only	200–204	122
Laying for elevation:		
Elevation counter	215–219	128
Gunner's quadrant	220–224	130
Materiel	240–244	136
Measuring:		
Angle of site to mask	230–234	132
Deflection	210–214	127
Elevation	225–229	131
Outline	194	119
Preparation	192	118
Purpose and scope	187	117
Qualification scores	193	119
Scoring	191	118
Sighting and fire control equipment	235–239	133
Standards of precision	188	117
Time	190	118
Telescope mount, T179E2:		
Cross level	153	98
Elevation counter synchronization	155	98
Longitudinal level	154	98
Training:		
Conduct	185	114
Objectives	184	114
Standards	186	115
Unloading the piece	125	77

[AG 472 (2 May 57)]

By Order of *Wilber M. Brucker*, Secretary of the Army:

MAXWELL D. TAYLOR,
General, United States Army,
Chief of Staff

Official:
HERBERT M. JONES,
Major General, United States Army,
The Adjutant General.

Distribution:
Active Army:
CNGB
Tec Svc, DA
Tec Svc Bd
Hq CONARC
Army Air Def Comd
OS Maj Comd
OS Base Comd
Log Comd
MDW
Armies
Corps
Div
Div Arty Brig
FA Gp
USMA
Gen & Br Svc Sch
Ord PG

Units org under fol TOE:
6–416, Hq&Hq Btry, FA Bn, 155-mm Gun or 8-in. How or 240-mm How or 8-in. Gun, Towed or Self-propelled
6–435, FA Bn, 155-mm Gun or 8-in. How, Self-propelled
6–437, FA Btry, 155-mm Gun, Self-propelled
6–447, FA Btry, 8-in. How, Self-propelled

NG: State AG; TOE: 6–416; 6–435; 6–437; 6–447.

USAR: TOE: 6–416; 6–435; 6–437; 6–447.

For explanation of abbreviations used, see AR 320–50.

Table I. *Duties in Preparing for Action*

Sequence	Chief of section	Gunner	No. 1	No. 2	No. 3	No. 4	No. 5	Driver, motor carriage	Ammunition specialist	Ammunition handler (2)	Driver, section vehicle
1	Commands PREPARE FOR ACTION. Supervises operations throughout all sequences.	Elevates tube to LOAD position after tube traveling lock is removed.	Unlock spade latches, unlock and open rear turret doors, and insert locking pins in braces after spade is lowered.		Stands by spade brake.	Lays wire for intrabattery communication system from motor carriage to executive's post.	Removes direct fire telescope and muzzle covers. Unlocks, lowers, and secures tube traveling lock.	Allows main engine to idle at 650 RPM.	Guides section vehicle into position for unloading ammunition.	Unload and prepare ammunition for firing, to include projectiles, powder charges, and fuzes, as directed by ammunition specialist.	Drives section vehicle in to position indicated by ammunition specialist.
2		Procures and installs panoramic telescope, levels telescope mount, and stows travel insert in telescope case. Sets horizontal equilibrator.			Controls lowering of spade. Detaches spade cable from spade.		Assembles aiming posts and places them near right front of motor carriage.		Supervises unloading of ammunition and its preparation for firing.		Assists in unloading ammunition.
3	Directs driver to back vehicle until spade is flush with the ground and the spade stops are seated against the bail. Commands driver to cut engine.		Remove breech cover.		Positions machine gun, cal 50, in direction of probable area of responsibility for position defense.	Procures and spreads paulin to left rear of motor carriage.	Procures and assembles unloading rammer to staff sections.	When directed by chief of section, backs vehicle onto spade. At the command of the chief of section, stops engine, puts transmission in neutral, sets hand brake. Checks auxiliary generator.			
4	Checks replenisher gages, equilibrator, and recuperator pressures.	Lays for direction, if required.	Opens breech; examines tube, breechblock, primer vent, and gas check pad; and cleans and oils parts, as required.	Places breech cover to rear of motor carriage.	Lowers loading trough into position for loading.	Places primer equipment and fuze setters on paulin.	Gathers covers and places to left rear of motor carriage.				
5		Bore sights piece, if time permits.	Assists gunner in bore sighting, if so directed.	Moves ammunition hoist to operating position. (8-inch howitzer).	Receives swab bucket and swab from No. 1.	Assists gunner in bore sighting, if testing target is used.	Passes swab bucket and swab to No. 3.	Assists in unloading ammunition or performs other duties, as directed by chief of section.			
6				Receives primers from No. 5. Takes post.			Procures and passes primers to No. 2.				
7	Verifies that piece is prepared for action. Reports to executive, "Sir, No. (so-and-so) in order," or reports any defects that cannot be corrected without delay by the section.	Directs No. 5 in setting out aiming posts, when so directed by chief of section or executive. Takes post.	Takes post.	Switches on turret ventilator, when appropriate. Takes post.		Cleans and oils fuze setter, if necessary. Takes post.	Sets out aiming posts, when so directed by gunner. Takes post.	Takes post.	After ammunition is unloaded, directs disposition of section vehicle. Reports to chief of section the number and type of projectiles, powder charges, and fuzes on hand. Takes post.	Takes post.	After ammunition is unloaded, drives vehicle to point indicated by ammunition specialist, stops engine, and performs preventive maintenance service.

147

Table II. Duties in March Order

Sequence	Chief of section	Gunner	No. 1	No. 2	No. 3	No. 4	No. 5	Driver, motor carriage	Ammunition specialist	Ammunition handler (2)	Driver, section vehicle
1	Commands MARCH ORDER and inspects chamber to insure that piece is not loaded. Supervises members of the section throughout all sequences.	Returns tube to LOCK position after tube traveling lock has been raised to upright position.	Secure projectile rack.	Secures propellant rack and receives breech cover from No. 5.	Attaches spade hoist cable to the spade.	Picks up wire and stows communication equipment.	Passes breech cover to No. 2. Raises tube traveling lock, secures tube, and replaces muzzle and direct fire telescope covers.	Starts engine. Sets hand throttle for warmup period at 1,000 RPM. Checks gages for proper functioning.	Directs section vehicle driver into position convenient for loading.	Loads ammunition or performs such duties as chief of section may direct.	Drives section vehicle to position indicated by ammunition specialist.
2		Moves auxiliary fire control panel switches to OFF position.	Insures that hydraulic rammer is in OFF position and that the rammer is in the retracted or traveling position. Replace breech cover.	Passes swab bucket, swab and primers to No. 5.	Moves ammunition trough and ammunition hoist (8-inch howitzer) to traveling position. Operates hoist controls and raises spade to traveling position.		Receives swab bucket, swab, and primers from No. 2. Recovers and stows aiming posts.		Supervises loading of ammunition.		Assists in loading ammunition.
3	Directs motor carriage driver to move carriage forward to unseat spade.	Removes and stows panoramic telescope and installs travel insert on telescope mount.					Recovers, disassembles, and stows unloading rammer.	At direction of chief of section, moves motor carriage forward to unseat spade.			
4		Insures that all telescope lights are off.	Close and lock rear turret doors.								
5	Verifies that piece is prepared for traveling.	Assists chief of section in checking motor carriage and section equipment.			Secure machine gun, cal .50, in traveling position.	Secure pioneer equipment and fuze setters. Assist in loading ammunition. Fold paulin and replace on motor carriage.					
6	Reports to executive, "Sir, No. (so-and-so) in order", or reports any defects that section cannot remedy without delay.					Take posts.					

148

Table III. Duties in Firing

Sequence	Chief of section	Gunner	No. 1	No. 2	No. 3	No. 4	No. 5	Driver, motor carriage	Ammunition specialist	Ammunition handler (2)	Driver, section vehicle
1	Directs work of section personnel throughout all sequences.	After auxiliary generator is started, turns on hydraulic pump motor and puts traversing-hydraulic motor shutoff valve in TRAV position. Lays weapon on commanded deflection.	Opens breech. Wipes chamber dry for first round, assisted by No. 3.	Moves ammunition hoist into position for receiving projectile (8-inch howitzer only).	Assists No. 1 in wiping chamber dry for first round. Inspects bore; if clear, announces BORE CLEAR.	Fuzes projectiles and sets fuzes. Assists No. 2 in connecting shot tongs to projectile and raising projectile with ammunition hoist (8-inch howitzer only).	Prepares propellant and hands it to No. 3, when requested.	Turns on master relay switch and starts auxiliary generator. Performs other duties as prescribed by the chief of section.	Supervises preparation of ammunition. Insures that adequate amounts of ammunition are prepared.	Prepare ammunition under the direction of the ammunition specialist.	Performs duties as prescribed by the ammunition specialist.
2, 3 (8-inch howitzer only).		Elevates or depresses piece to loading elevation. Turns on scavenging compressor.		Raise projectile with ammunition hoist and places it on loading trough against rammer chain head.	Lowers loading trough into breech after piece is at loading elevation.	Assisted by ammo handlers, carry projectile to rear of motor carriage.				Assist in carrying projectile to rear of motor carriage.	
2 (155-mm gun only).			Receive projectile on loading tray from No. 4 and 5 and set it on lower rear door.			Raise projectile on loading tray and hand it to No. 1 and 2.					
3 (155-mm gun only).				Picks up projectile and places it on loading trough against rammer chain head.	Receives propellant from No. 5 and passes it to No. 2, when requested.						
4	After projectile is in correct position on rammer trough, signals "Ram."		At command of chief of section, rams projectile.	Receives propellant from No. 3.	Raises loading trough.						
5	Insures that weapon is ready to fire. Indicates to executive that weapon is ready to fire by raising arm or by announcing orally.	Elevates weapon to commanded elevation. Announces (signals) READY after No. 1 has announced SET.	Closes breech after loading trough is raised and after No. 2 has announced CLOSE. Announces SET after No. 2 has announced LOADED.	Inserts propellant into chamber and announces CLOSE. Inserts primer into firing lock after breech is closed and announces LOADED.							
6	On command of executive, commands FIRE to gunner and/or lowers arm.	Fires piece as command of chief of section.	Moves safety fire switch to ON position after gunner announces (signals) READY.								
7		Elevates or depresses piece to loading elevation.	Opens breech after primer is removed.	Removes spent primer.	Swabs powder chamber after each round. Cleans obturator spindle vent and wipes mushroom head, when necessary. Inspects bore; if clear, announces BORE CLEAR.						

Table VI. *Duties in Inspection and Maintenance*
Before Operation (B) and During Halts (D)

Sequence	Chief of section	Gunner	No. 1	No. 2	No. 3	No. 4	No. 5	Driver, motor carriage	Ammunition specialist	Ammunition handlers (2)	Driver, section vehicle
1	Supervises inspection by all members in all sequences.	(B) (D) Inspects for condition, functioning, and security of sighting and fire control equipment.	(B) Assisted by No. 2, unfastens and removes breech cover.	(B) Assists No. 1 in unfastening breech cover. Inspects for tears, wear, and broken or missing fastenings.	(B) Inspects fixed fire extinguisher.	(B) Inspects communication equipment. (D) Inspects track on left side for condition.	(B) (D) Inspects tracks on right side for condition and tension.	(B) (D) Fills out vehicle operation record. Inspects publications and Standard Form 91.	(B) Supervises loading and checking of ammunition and inspection services by ammunition handlers and section vehicle driver.	(B) (D) Inspects ammunition and loads personal equipment.	(B) (D) Fills out vehicle operation record. Inspects publications and Standard Form 91.
2	(B) (D) Cylinder gages to determine quantity of recoil oil and if directs No. 2 to service the system, if necessary.	(B) Tests operation of traversing and elevating mechanism (power and manual).	(B) Opens breech and checks for completeness and condition of mechanism.	(B) Checks cal .50 machine gun ammunition hoist (8-inch howitzer) and powder stowage rack for security.	(B) Checks cal .30 machine gun and fuze setters for cleanliness, functioning, and security. Checks stowage of machine gun ammunition.	(B) Cleans and oils fuze setters. (D) Inspects bumper springs, road wheel arms, shock absorbers, and torsion bars on left side for condition.	(B) (D) Inspects bumper springs, road wheel arms, shock absorbers, and torsion bars on right side for condition.	(B) (D) Checks fuel and oil levels, looks for leaks in engine compartment.	(D) Inspects ammunition for security.	(D) Inspects and removes all foreign matter such as nails, glass, and stones from tread.	(B) (D) Checks fuel in tank, notes leaks, and adds fuel. Checks oil level and adds oil, if necessary. Checks coolant level and notes any leaks. Checks tires and security of spare wheels and tires.
3	(B) Verifies presence of all technical manuals, lubrication orders, trip tickets, driver's licenses, and accident report form.	(B) Checks functioning of night lighting devices.	(B) (D) Checks projectile stowage rack for security.	(B) Services recoil system, if directed by chief of section.	(B) Checks operation and security of spade controls and spade. (D) Checks spare containers for contents. Assists driver in filling fuel tank.	(B) Checks operation of bilge pump and ventilating blower.	(B) (D) Checks ground under vehicle for leaks.	(B) (D) Checks spare containers for contents.		(B) (D) Checks under truck for indication of fuel, oil, coolant, gear oil, and brake fluid leaks.	(B) (D) Visually inspects lamps, reflectors, horn, fire extinguisher, mirrors, pauline, tools, etc., to determine if they are in the proper place and in good operating condition.
4	(B) Verifies proper supply of gasoline, oil, water, and emergency rations.	(B) Verifies completeness of section equipment and cleaning and preserving material.	(B) Checks operation of fire warning light and power rammer.	(B) Inspects piece and mount for loose parts and cracked or broken welds.		(B) Assist gunner in checking completeness of section equipment.	(B) Assist driver in checking condition and operation of lights.	(B) (D) Assisted by No. 5, checks condition and operation of lights.			(B) Starts engine and checks all instruments for normal readings.
5	(B) Checks section equipment for proper loading and completeness.	(B) Bore sights the piece. (Bore sighting should be checked after displacement and prior to firing, if time permits.)	(B) Assists gunner in bore sighting.			(B) Assist gunner in bore sighting.	(B) Assist gunner in bore sighting. Restores testing target.	(B) Starts motor. Observes instruments for normal readings during warmup.			
6	Reports to battery executive, "Sir, No. (so-and-so) in order," or reports any defects which the section cannot remedy without delay.	Reports, "Gunner ready."	Reports, "No. 1 ready."	Reports, "No. 2 ready."	Reports, "No. 3 ready."	Reports, "No. 4 ready."	Reports, "No. 5 ready."	Reports, "Driver, motor carriage, ready."	Reports, "Ammunition section ready."		

After Operation

Sequence	Chief of section	Gunner	No. 1	No. 2	No. 3	No. 4	No. 5	Driver, motor carriage	Ammunition specialist	Ammunition handlers (2)	Driver, section vehicle
1	Supervises maintenance and inspection by all members in all sequences.	Cleans and tests sighting and fire control equipment.	Cleans and lubricates breech mechanism, assisted by No. 2.	Assists No. 1 in cleaning breech mechanism.	Cleans and lubricates tube, assisted by No. 4 and the two ammunition handlers.	Assists No. 3 in cleaning the tube.	Inspects tracks for condition and tension.	Idles engine properly before stopping. Observes instrument and warning light while engine idles. Shuts off master switch. Checks fuel and oil levels. Looks for leaks in engine compartment. Checks spare containers for contents. Checks fuel fillers for leaks. Operates lights and horn (if tactical situation permits.)	Inspects, cleans, and lubricates cal .50 machine gun. Supervises maintenance of section vehicle and equipment.	Assists No. 3 in cleaning the tube.	
2	Cleans and tests gunner's quadrant.	Cleans and tests traversing and elevating mechanism.	Inspects, cleans, and lubricates cal .50 machine gun, assisted by No. 2.	Assists No. 1 in cleaning cal .30 machine gun.			Inspects bumper springs, road wheel supporting arms, shock absorbers, road wheels, and track supporting rollers.	Cleans battery, checks water level, and inspects terminals for corrosion, tightness, and coating of grease. Completes vehicle operation record.			
3	Inspects recoil system and directs No. 2 to service, if necessary.	Assists chief of section in supervising maintenance by all members of the section.	Cleans, lubricates, and tests firing lock.	Cleans inside of turret.	Cleans and stores rammer staff, aiming posts, and pioneer tools.	Inspects and cleans communication equipment.			Performs other duties as indicated above.		
4	Posts weapon record books and verifies presence of all forms and manuals.		Checks condition of fixed fire extinguisher.	Services recoil system, if necessary.	Inspects and cleans section tools and tool stowage compartment.	Cleans vision devices.	Visually inspects hull, howitzer traveling lock, turret, towing connections, and hull doors. Lubricates items specified on lubrication chart.				
5	Reports to battery executive, "Sir, No. (so-and-so) in order," or reports any defects that cannot be corrected without delay.										

150

IN HIGH DEFINITION NOW AVAILABLE!

COMPLETE LINE OF WWII AIRCRAFT FLIGHT MANUALS

WWW.PERISCOPEFILM.COM

©2013 Periscope Film LLC
All Rights Reserved
ISBN#978-1-937684-51-8
www.PeriscopeFilm.com

www.ingramcontent.com/pod-product-compliance
Lightning Source LLC
LaVergne TN
LVHW051836080426
835512LV00018B/2918